This collection of essays offers [...] rary feminist practices, with pa [...] politics and poetics of space. W[ith the growing importance of] technology and communications, the effects of globalisation and the change in social demands, feminist praxis of space is diversifying and expanding. This book assesses these *altering practices* alongside the emergence of new social and political theories and within today's urban and geopolitical contexts. The practices discussed in this book constitute a multitude of gestures, figures and actions: 'urban curating', 'making space', 'taking place', 'urban cooking', 'building while being in it', 'drifting walls', 'mapping invisible privileges', 'confessional constructions', 'stray sods'. Together they form a piece of feminine textuality, a volume which is not divided into thematic sections but takes its shape from a 'mise en relation entre des voix-femmes' in which each contribution negotiates its proximity to the others.

The contributions from architects, artists, cultural theorists and activists from different generations and a wide range of practices and fields of research are international in scope and combine to challenge traditional disciplinary and cultural boundaries.

Doina Petrescu is senior lecturer in architecture at the University of Sheffield. She has written, lectured and practised individually and collectively on issues of gender, technology, (geo)politics and poetics of space. She is co-editor of *Architecture and Participation* (London: Routledge, 2005).

ALTERING PRACTICES

Feminist
Politics and Poetics
of
Space

Edited by **Doina Petrescu**

Routledge
Taylor & Francis Group
LONDON AND NEW YORK

First published 2007 by Routledge
2 Park Square, Milton Park, Abingdon, Oxon, OX14 4RN

Simultaneously published in the USA and Canada by Routledge
270 Madison Avenue, New York, NY10016

Routledge is an imprint of the Taylor & Francis Group,
an informa business

© 2007 Doina Petrescu, selection and editorial matter, individual
chapters, the contributors

Design and typsetting by Bev Weaner and Annabel Fraser
Additional typesetting by Sandra Zellmer
Typeset in Parable, Celeste, Fedra Mono, Helvetica and Letter Gothic
Printed and bound in Great Britain by The Alden Press, Oxford

All rights reserved. No part of this book may be reprinted or
reproduced or utilised in any form or by any electronic, mechanical,
or other means, now known or hereafter invented, including photo-
copying and recording, or in any information storage or retrieval
system, without permission in writing from the publishers.

British Library Cataloguing in Publication Data
A catalogue record for this book is available from the British Library

Library of Congress Cataloging-in-Publication Data
Altering practices : feminist politics and poetics of space.
    p. cm.
  Includes bibliographical references.
  ISBN-13: 978-0-415-35785-2 (hardback: alk. paper)
  ISBN-13: 978-0-415-35786-9 (pbk: alk. paper) 1. Communication in
architecture. 2. Academic writing. 3. Space (Architecture). 4. Cities and
towns—Philosophy. 5. Critical theory.

NA2584.A48 2006
720.82—dc22

2006031233

ISBN10 0-415-35785-3 (hbk)
ISBN10 0-415-35786-1 (pbk)

ISBN13 978-0-415-35785-2 (hbk)
ISBN13 978-0-415-35786-9 (pbk)
ISBN13 978-0-203-00393-0 (ebk)

CONTENTS

	Notes on the contributors	vii
	Foreword: from Alterities and beyond	xiii
1	Altering practices Doina Petrescu	1
2	Taking place and altering it Teresa Hoskyns, Doina Petrescu and other mixed voices	15
3	Evaluating Matrix: notes from inside the collective Julia Dwyer and Anne Thorne	39
4	An invisible privilege muf	57
5	How to take place (but only for so long) Jane Rendell	69
6	Building while being in it: notes on drawing 'otherhow' Katie Lloyd Thomas	89
7	Stray Sods: eight dispositions on 'the feminine', space and writing Brigid McLeer	113
8	Micro-strategies of resistance Helen Stratford	125
9	Altering events in architecture Anne Querrien	141
10	Urban curating: a critical practice towards greater 'connectedness' Meike Schalk	153
11	Open Kitchen or 'cookery architecture' Kim Trogal	167

12	Building clouds, drifting walls Ruth Morrow	189
13	Urban traces: civic performance art and memory in public space Ilana Salama Ortar	201
14	Sex & Space: space / gender / economy Marion von Osten	213
15	Refiguring dis/embodiments Niran Abbas	241
16	Stabat Mater: on standing in for matter Francesca Hughes	259
17	The unbearable being of lightness Jennifer Bloomer	281
18	Learning and building in the feminine Sadie Plant	297

NOTES ON THE CONTRIBUTORS

Niran Abbas is a senior lecturer in media and cultural studies at Kingston University. She is the editor of *Mapping Michel Serres* (2005) and *Thomas Pynchon: Reading from the Margins* (2003) and author of *The Discourse of Artificial Intelligence* (2006). Her interests are in new media, surveillance, the transhuman and digital media. Currently she is working on *The Five Senses and New Technologies*.

Jennifer Bloomer is a principal of JB/RS Architecture Design Planning in Athens, Georgia, USA. She is the author of *Architecture and Text: The (S)crypts of Joyce and Piranesi* (1993, 1995) and of four dozen published essays in English, French, Swedish, and Slovak.

Julia Dwyer practises across art and architecture, combining architectural practice with collaborative public art projects. She lectures at the University of Brighton, and is an associate lecturer at Chelsea College of Art and Design.

Teresa Hoskyns is an architectural designer, senior lecturer at Brighton University and currently working on a part-time PhD by practice at the Barlett School of Architecture, UCL, on the general theme of democracy and public space. Her publications include: 'Taking Place', *Public Art Journal* (2000), 'The Empty Place of Power', *Scroope* (2002), 'Designing the Agon: Questions on Space, Architecture and the Political', in Bruno Latour and Peter Weibel (eds) *Making Things Public* (2005), 'City/Democracy: Retrieving Citizenship', in Peter Blundell Jones, Doina Petrescu and Jeremy Till (eds) *Architecture and Participation* (2005).

Francesca Hughes is a critic and educator, teaching at the Architectural Association, London. She collaborates with artist Jonathan Meyer in a multi-disciplinary practice that incorporates architecture and the visual arts. She is the editor of *The Architect: Reconstructing Her Practice* (1996), and has written and lectured extensively. Her forthcoming book is *The Architecture of Error*.

Katie Lloyd Thomas is currently completing a PhD about language and materiality at Middlesex University. She has been working in architectural education for more than ten years, most recently at the University of East London. She researches and writes on feminism, architectural representation and materiality and is editor of *Material Matters: Architecture and Material Practice* (2006). She has given talks in France, Iceland, Slovakia, Australia, Germany and the UK and is a founder of 'Taking Place', a group of architects, artists and writers working with questions of feminism and space. She is collaborating with Brigid McLeer on 'In Place of the Page', an ongoing project which explores the space of the page through a series of translations from emails to textplans to architectural drawings, which has been exhibited in London, Nottingham, Brighton and Drogheda, Ireland.

Brigid McLeer is an Irish artist based in London. Her work is process-based and concerned with questions of translation and location. Recent exhibitions include: 'Spatial Imagination' at Domo Baal Gallery, London and 'In Place of the Page' at Standpoint and Unit 2 Galleries, London, Phoenix Gallery, Brighton and Bridport Arts Centre. Her recent publications include: *Spatial Imagination*, exhibition catalogue (2006), a chapter in Sharon Kivland, Jaspar Joseph-Lester and Emma Cocker (eds) *Transmission: Speaking and Listening* (2005) and 'Moving Bodies', in Ric Allsop, Richard Gough, Claire MacDonald (eds) *Performance Research* (2003). She is currently a visiting lecturer in fine art at Goldsmith's College and Oxford Brookes University and in history and theory at the Bartlett School of Architecture, UCL.

Ruth Morrow graduated from Dundee School of Architecture in 1989. She practised architecture in Belfast and Berlin and has taught at several schools of architecture throughout Britain and Ireland. In 2003 she was appointed Professor of Architecture in the School of Art and Design at the University of Ulster. Her research crosses two areas: inclusive design and the process of design and its associated pedagogy. She uses an activist agenda as a place in which to test and integrate inclusive values and creative skills.

muf architecture/art were established in 1994 with the then eccentric desire to make work in and for the public realm. They have been doing so since. They have published *This Is What We Do: A muf Manual* (2001).

Ilana Salama Ortar is an art curator and artist working in the field of *civic performance art* including: interdisciplinary art interventions dealing with memory and alternative proposals for commemoration in public space; art projects opening up public discussions on the subject of political-social urgency, the condition of immigration, refugee, erased memory and identity loss within the urban realm. She is currently completing a PhD at the Roehampton University, Surrey, on civic performance art in relation to the architecture of emergency and population displacement. Her main exhibitions are: *The Visible and the Invisible in Israeli and Palestinian Memory* (installation/performance), Israel Museum, Jerusalem (1994), *The Prophets' Tower*, Shopping Center, Haifa (1995), *Uprootedness, Refugee-ism and Immigration through the Grand Arenas Transit Camp of Marseilles*, Museum of Contemporary Art, Marseilles and Herzeliya Museum of Contemporary Art, Israel (1998-2005). Her last book and documentary film are *The Camp of the Jews, Uprooting, Refugeeism and Immigration* (2005).

Doina Petrescu is an architect, urban curator and writer, currently teaching at the University of Sheffield, and at the Architectural Association, London. She has practised, taught and lectured extensively on issues of gender, technology, politics and poetics of space and been an activist with local associations in the UK, Germany, France, Romania and Senegal and feminist research groups such as 'association (des pas)' in Paris and 'Taking Place' in London. She is a co-founder of *atelier d'architecture autogérée (aaa)*, a networked practice including architects, artists, activists, researchers and residents aiming to develop critical 'tactics' in architecture and urban planning. She is a co-editor with Jeremy Till and Peter Blundell Jones of *Architecture and Participation* (2005) and co-author with *aaa* of a forthcoming book on alternative practice in architecture.

Sadie Plant is the author of *The Most Radical Gesture: The Situationist International in a Postmodern Age* (1992), *Zeros and Ones: Digital Women and the New Technoculture London* (1997) and *Writing on Drugs* (1999). She has been a lecturer in cultural studies at the University of Birmingham, and Research Fellow and Director of the 'Cybernetic Culture Research Unit' at the University of Warwick. She left the academy in 1997, and since then has been writing, travelling, and speaking at events around the world. She is currently working as a freelance author on the topic 'social and cultural implications of new technologies'.

Anne Querrien is a sociologist and urban planner. In May 1968 she was an activist in the *22 March Movement* and in the 1970s she worked as a researcher at CERFI (Centre d'études, de recherches et de formations institutionnelles) with Félix Guattari. She is an editor of *Annales de la Recherche Urbaine* and member of the editorial committee for the journals *Multitudes* and *Chimères.* Parallely, she activates in different associations such as AITEC (Association Internationale de Techniciens, Experts et Chercheurs) and CLCV (Consomation, Logement et Cadre de Vie).

Jane Rendell is a Reader in Architecture and Art and Director of Architectural Research at the Bartlett School of Architecture, UCL. An architectural designer and historian, art critic and writer, she is engaged in interdisciplinary research, and author of *Art and Architecture: A Place Between* (forthcoming, 2006) and *The Pursuit of Pleasure* (2002), editor of 'Critical Architecture', a special issue of the *Journal of Architecture* (June 2005), and co-editor of *Strangely Familiar* (1995), *Gender, Space, Architecture* (1999), *Intersections* (2000) and *The Unknown City* (2000), and *Spatial Imagination* (2005).

Meike Schalk is an architect educated at the Hochschule der Künste in Berlin and at the Architectural Association in London and is currently completing a PhD in landscape planning in Sweden. She has worked in architectural practices in Stuttgart, Berlin, London and Stockholm since 1986, and taught architectural and urban design 1997-2003 at the Royal Institute of Technology in Stockholm. She is an editor for the cultural magazine *SITE*. In her work, research and practice converge as with the collaborative platform *urban curating* (2000–3) in various projects. Her main interests include social dynamics on the micro-scale, questions of local cultural production, alternative economies, and participation in planning.

Helen Stratford is an architect. Her practice, located between architecture, writing and performance, has formed the basis of exhibitions and presentations at international conferences and events, including Tate Modern, the Architecture Foundation London, Columbia University, New York, the Ecole Supérieure des Beaux Arts, Paris, and the Living Art Museum, Reykjavik. She was a Resident Fellow in Architecture at Akademie Schloss Solitude, Stuttgart in 2004 and a member of the feminist group 'Taking Place', and has worked at MOLE and 5th studio, Cambridge. She was assistant curator of the *Compendium* exhibition, RIBA, London, 2006.

Anne Thorne is a partner in an all-women practice which designs for sustainable communities and a sustainable environment, working in participation with clients, users and artists. Her projects include urban planning, social housing, nurseries, respite care and community buildings.

Kim Trogal graduated from Sheffield University and worked at *fluid*, a multi-disciplinary architectural practice and think tank. With *fluid* she has taught a postgraduate diploma unit at London Metropolitan University and has explored dialogue, participation and alternative strategies at the level of both master planning and architectural projects. She currently works in private practice and is preparing to undertake a PhD, exploring 'feminine' practices in architecture and urban planning.

Marion von Osten is an artist, curator and writer based in Berlin. Her projects are primarily concerned with the changed conditions in which cultural work is produced in neo-liberal societies, technologies of the self and the governance of mobility. Between 1999 and 2006 she has taught at the Institute for the Theory of Art and Design and the Institute for Cultural and Gender Studies, HGK Zurich. Since 2006 she has held a Professorship at the Academy of Fine Arts, Vienna. Between 1996 and 1998 she was a curator at Schedhalle, Zurich and since then has initiated a number of exhibitions and research projects including *Transit Migration*, *Be Creative! The Creative Imperative*, *MoneyNations*, *Sex & Space* and *SuperMarket*. She has edited (with Justin Hoffmann) *Das Phantom sucht seinen Mörder* (1999) *Norm der Abweichung* (2003) and *Projekt Migration* (2005) and has published in *Kultur der Arbeit* (2003), *Gender und Globalisation* (2001) and *Agenda: Perspectiven kritischer Kunst* (2000).

Doina Petrescu

FOREWORD:
from Alterities and beyond

Maria Puig de la Bellacasa

Foreword   Doina Petrescu

Jane Rendell

Sadie Plant

Rosi Braidotti

Altering practices was initially called *Alterities*: its genesis was a conference in Paris, a manifestation that wanted to record the recent production of a feminism of difference and diversity within the practice of space.[1]

I organised this conference at the time because I was interested in bringing together different genealogical lines within the feminist approach to architecture in the late 1990s. I was interested in giving an account about 'where we were' at that moment and to mediate an encounter between theorists from North America (such as the group that produced the issue of the *ANY* magazine on 'Architecture and the Feminine'), feminist practices and theories from United Kingdom (such as the Feminist Design Collective, Matrix, muf and the authors of books such as *Making Space, Reconstructing Her Practice, Desiring Practices* or *Gender, Space and Architecture*), and a few French interdisciplinary practices, more or less rooted in what is commonly called *French feminism*. People involved in these undertakings were invited to *Alterities*, together with other feminists from different fields who had responded to an open call for papers. We tried to identify politics and poetics within the relationship between art and architecture, including issues of technology, bodies and space, gendered practices and situated knowledge.

The organisation of *Alterities* was made possible through a series of coincidental networks and spatial and temporal opportunities: Sylvie Clavel, as head of the school of architecture at Paris Villemin and Mathilde Ferer, a feminist in the Ecole des Beaux Arts administrative staff – which were two strategic positions within the institutions that jointly hosted the conference – and also Jennifer Bloomer, Robert Segrest, Karen Bermann, Pat Potter and Catherine Ingraham as supporting colleagues at the University of Iowa, where I was

1 *Alterities* was co-organised by l'Ecole d'Architecture Paris Villemin and l'Ecole Nationale Supérieure des Beaux Arts in Paris, 4–5 June 1999. Participants in the conference included: Niran Abbas, Karen Bermann, Jennifer Bloomer, Martine Bouchier, Christine Buci-Glucksman, Raoul Bunschoten, Mireille Calle-Gruber, Jean-François Chévrier, Harris Dimitropoulos, Sonja Diquemare, Mary Flanagan, Christian Girard, Francesca Hughes, Catherine Ingraham, Jecca, Olga Kiseleva, Tina Laporta, Neil Leach, Bracha Lichtenberg-Ettinger, Katie Lloyd Thomas, Vera Mantero, Petra Marguc, Matrix (Julia Dwyer and Anne Thorne), Brigid McLeer, Fiona Meadows, Monique Minaca, muf, Sally Munt, Ilana Salama Ortar, Marie-Paule Pages, Doina Petrescu, Sadie Plant, Patricia Boinest Potter, Anne Querrien, Jane Rendell, Victoria Rosner, Carolsan Shea, Lucy Sheerman, Helen Stratford and Andrea Wheeler.

a guest professor at the time, and Hélène Cixous as my PhD supervisor in the department of Etudes Féminines at the University Paris 8. Without them, the conference would never have happened in a location and context that were not particularly open to feminism at the time. *Alterities* was a significant feminist event but also an occasion to confront positions and perspectives, to empathise, to connect and go further. It has also empowered some of us to continue, restart, revise and eventually reinvent our practices.

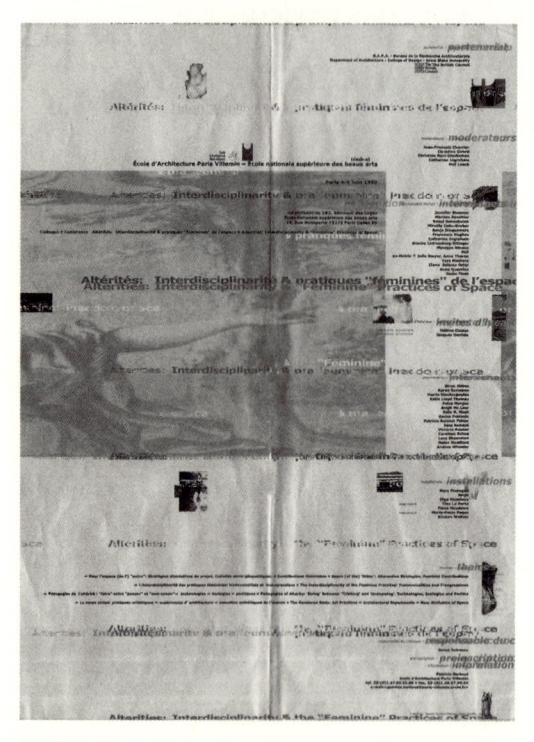

Alterities poster

I realise now that writing about feminist practices in architecture, twelve years after the publication of *Architecture and the Feminine: Mop-Up Work*, when issues of feminism in architecture looked already outdated, still remains a marginal undertaking. Jennifer Bloomer had noted even then that 'the surface has been barely scratched in this area. There must be 40 years of work to be done on architecture and the feminine.' This 'scratched surface' she refers to nevertheless does metaphorically constitute a ground. Not a ground to settle on, but a ground to allow growth, to encourage others to go further in their practices.

I organised *Alterities* because I wanted to continue the commitment for the next '40 years' with the 'mop-up work', the 'reinvention of her practice', the 'making of space', the 'nomadic subjectivity' in architecture. The poster that we designed for the conference reinterpreted the *ANY* publication cover and the word 'feminine' was put in inverted commas in the title to signify the critical use of a term that had been taken apart from its meaning and history and split into new discussions.[2] The title of the conference reiterated the question of the 'feminine' from the critical stance of the 1990s and the perspective of current feminist theories. The use of inverted commas indicated that we were no longer speaking of 'woman' and her spatial practices within a theory of dichotomy and a dream of unity, but more within a heterogeneous spectrum of the 'feminine' coming under a theory of 'alterity'.

'Alterities' became an invented word to name the multiple possibilities of praxis: 'other spatial practices' or practising 'otherwise', expressing alternative and alterative positions formulated according to the current re-compositions of individual and collective subjectivities within the new technological and geopolitical contexts. We were also speaking 'in several languages' from non-exclusive and non-generalising positions, trying to overcome the binary of traditional logics and politics and undermine conventional institutions and dominant ideologies. The conference addressed different fields and disciplines: architecture, art, media, psychoanalysis, literature, sociology, politics.[3] We were aware that translation and trans-disciplinary moves are familiar to any feminist approach because women and relations between genders are everywhere and cannot be contained within a single disciplinary framework.

---

2   The discussion panels in *Alterities* included: Spaces (of the) 'Other': Alternative Strategies of the Architectural Project. Geopolitical Crossings. Feminist Contributions. The Interdisciplinarity of the Feminine Practices: Transversalities and Transgressions. Pedagogies of Alterity: 'Doing' Between 'Thinking' and 'Unknowing'. Technologies, Ecologies and Poetics. The Gendered Body: Art Practices – Architectural Experiments – New Aesthetics of Space.

3   *Alterities* pointed to what we could call a 'situated trans-disciplinarity' – respectful of closures, especially when we are willing to transgress them.

Instead of 'architecture', we took 'space' as a more open category, theorised within a trans-disciplinary framework.[4] The contemporary understanding of 'space' has changed from its traditional metaphysical roots. As Sadie Plant notes in her chapter, the contemporary paradigms and rules emerging from technology and cyberspace contaminate the traditional concept of space, forcing a thinking of it as a fluid environment composed of self-regulating, reactive and intelligent networks. An architecture which integrates intelligent materials and self-regulatory systems must be conceived, according to Sadie, as a process which generates continuous change. It is 'space' which represents, in some way, contemporary 'alterity'.[5]

Improvision: 5 minutes. Performance by Vera Mantero in Palais des Etudes of Ecole Nationale Supérieure des Beaux Arts, 4th June 1999 (Credit: D. Petrescu)

The understanding of space has also changed with the evolution of global processes, the current social conflicts and their geopolitical reconfigurations. Space is today understood differently by the feminist theorists who are increasingly concerned with issues of mobility, 'nomadism', 'positioning' and 'situatedness'. 'Connectionist models' mark also a shift away from the central, organising, transcendent thinking of evolutionary processes within space and architecture.

The excitement after the conference generated the idea of a collectively edited book, but as Jennifer said at the time, a book has to be thought of not as a solid object but as a bag that takes the shape of what is contained. The 'bag' filled with

---

4   In her co-edited book, *Gender, Space, Architecture*, Jane Rendell talked about 'space' as the topic which allowed the start of an interdisciplinary feminist approach to architecture.

5   See Sadie Plant's Chapter 18 in this volume.

Alterities papers has since changed shape, it has been altered by time. Some of the chapters in this book relate back to *Alterities*, others are newly invited contributions, trying to update the problematic that was discussed at the conference but also trying to mark a shift in time. Having to publish today a book that originated in this event and having to transfer papers initially submitted on floppy discs to DVDs, I felt it was important at the same time to tell the story of *Alterities*, to reveal its context within feminist theory and practice in the late 1990s and what has happened since. Seven years later, the same story is told differently, from an already historical perspective, in which 'history' is both that of the feminist practices (of space) and the intersection between the 'hi/stories' and 'herstories' of some of those involved in the conference.[6] All these 'collective fabulations' that 'build in the past in a way that challenges the future', all these accounts

6   According to Maria Puig de la Bellacasa: *Writing hi/stories instead of histories or stories is a small attempt to deviate from the truth or fiction alternative. The accounts we make through our knowledge constructions are rooted in reality and representations. They are not mere illusionary fictions nor mere real truths, they are collective fabulations building the past in a way that challenges the future. Fabulation is a concept that I take from G. Deleuze and F. Guattari. Fabulation is a constitutive gesture of a community, which marks the creative character of truth.*

'Feminist knowledge politics in situated zones: A different hi/story of knowledge construction' (*womeninweb.women.it/cyberarchive/files/puig.htm*)

Alterities discussion (Credit: D. Petrescu)

disseminated in the papers, constitute a multi-voiced narrative, a multiplicity of personal and creative overviews of the event, that I wanted to record as well. I have also tried to situate this book within the context of its production, knowing from Braidotti that '*situatedness* is a powerful appeal to yearning, an effort to work out shared meanings of a situation without trying to transcend it. A situated account of knowledge construction resembles more an implicated mapping gesture than a normative foundational gesture.'[7] I wanted *Altering practices* to be a situated account, an implicated mapping gesture of what happened before and after *Alterities*.

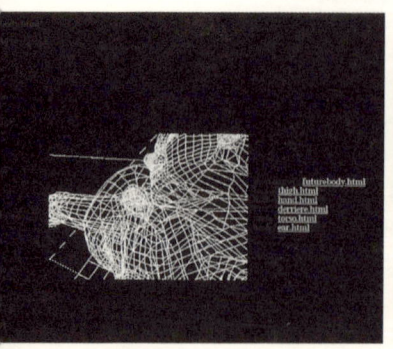

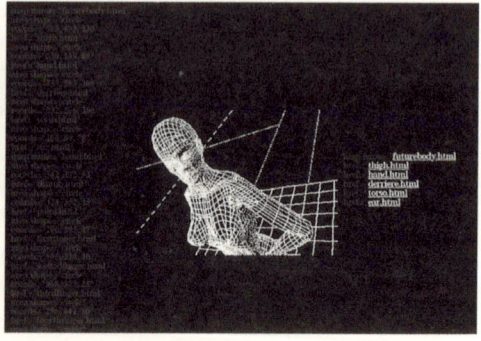

Future_body, version 1.0 (Credit: Tina Laporta)

---

[7] Rosi Braidotti, *Nomadic Subjects*, New York: Columbia University Press, 1994.

Chapter 1

Maria Puig de la Bellacasa
Doina Petrescu

ALTERING PRAC

Rosi Braidotti

Judith Butler

Geraldine Pratt and Pauline Hanson

Chapter 1   Doina Petrescu

bell hooks

TICES

Jacques Rancière

Becomings, yearnings

The *Altering practices* based their meaning on *Alterities*. They both refer to *alter* – the Latin word for 'other' – more as a verb than a noun. They speak about making or becoming different, about change, a change that could relate to time or directly to gender.[1] In a Deleuzian language, *Altering practices* are 'becomings'; they are 'active, dynamic processes of thinking and transformation, and an affirmation of "difference" as a positive quality'. This notion of 'becoming' has already been appropriated by the contemporary feminist project. Feminist thinkers such as Rosi Braidotti or Elisabeth Grosz have theorised the feminist subject as 'a term in a process of intersecting forces [affects], spatio-temporal variables that are characterized by their mobility, changeability, and transitory nature'.[2] *Becomings* are 'ways of establishing concrete material and semiotic connections among subjects that are conceived in terms of a multiplicity of forces/affects'.[3]

While the logic of *'becoming'* may offer the potential for an infinite variety of constellations, forming and reforming in perpetual change, specific *'becomings'* are always located, they are always fostered by their particular situation, historically, materially and critically.

Understood in these terms, the present book maps a particularly located and materialised transformation of the contemporary feminist project in architecture but also, as Francesca Hughes notices in her chapter, a certain becoming of the architectural theory and practice in general.

The *Altering practices* are named after the process they induce, *by what they do*. They rely on the transformative power of 'altering', on both its positive and critical dynamics. In the syntagm 'altering practices', 'altering' could mean 'undermining', 'subverting' received identities and authoritative rules, norms and tools and working out other shared meanings throughout their transformation; it could also be an appeal to *yearning*. 'Yearning', this gift made to theory by the black

---

1 NB: To 'alter' in American English could literally mean 'to change gender'.

2 Rosi Braidotti, *Nomadic Subjects*, New York: Columbia University Press, 1994, p.111.

3 Ibid.

feminism, combines hope, moaning and desire, everything that a soul hungers for and is powerless to define, that which will transform the soul rather than be appropriated by it. It brings a sense of immanence that, instead of overcoming contradictions, enables us to slip through them. As bell hooks has put it, the question of 'yearning' is not 'who we are' but 'what we want to become'; the *Altering practices* are about *what we want the world to become...*[4]

Collective re-constructions

Altering practices have taken further the 'mop-up work' begun by their predecessors into a multiplicity of figurations, a number of gestures, figures and actions that constitute the spectrum of these practices in this volume: *'urban curating', 'making space', 'taking place', 'urban cooking', 'drifting walls', 'mapping invisible privileges', 'confessional constructions'...*

They are accompanied by a collection of concepts and metaphors: *chaos, complexity, fluidity, emergence, lightness, connectionism, multiplicity, networks, self-organisation...* We realised that most of the paradigms that structure our contemporary thinking of space have been shaped by the feminist imaginary. But most of them have been brought into theory and practice through technologically and scientifically rooted discourses, losing their poetical and political dimensions. Contributions to this book, such as those by Francesca Hughes, Jennifer Bloomer or Sadie Plant, try to re-appropriate such paradigms. They define the *Altering practices* as gestures of discursive re-appropriation of theoretical territories and actions that frame the contemporary understanding of space.

This act of re-appropriation within a feminist project has to do with knowledge construction and knowledge politics. A political approach to knowledge means, in feminist terms:

a delicate and implicated re-construction of its processes, the search of the links and relations between many different elements often irreducible one to another. In feminism, this gaze

---

[4] bell hooks, *Yearning: Race, Gender, and Cultural Politics*, Boston, MA: South End Press, 1990.

into knowledge is informed by confidence on *collective* re-constructions, not only on theoretical deconstructions.[5]

These *collective reconstructions*, in our case, suppose ways of doing and undoing, ways of making and remaking space, of 'producing space' according to 'altered' rules and values. These ways are both political and poetic.

We can also talk about *Altering practices* as practices of 'curating', of 'care taking', acknowledging the work of reconstruction and re-production that bears ethical and emotional charge, the kind of work that was always associated with women. Authors in this volume talk about the importance of memory and genealogies in practices of 'urban curating' (Meike Schalk), 'making space' (Matrix), 'urban cooking' (Kim Trogal), 'civic performance' (Ilana Salama Ortar), 'detailing' (Helen Stratford), 'drawing "otherhow"' (Katie Lloyd Thomas), 'mapping invisible privileges' (muf), 'confessional constructions' (Jane Rendell), 'counter-representations of difference' (Marion von Osten), 'stepping on *stray sods*' (Brigid McLeer), 'longing' and 'standing in for matter/mater' (Jennifer Bloomer and Francesca Hughes), 'building clouds and drifting walls' (Ruth Morrow).

Otherhow

These reconstructions also affect the subjects themselves, constituting at the same time, as Butler has argued, subversions of identitary constructions. The *Altering practices* are altering identities and re-territorialising domains that no longer correspond to traditional categorisations. Chapters in this volume reveal different kinds of practices of writing, teaching, building, planning or art that are breaking barriers between theories and practices, academia and activism and enabling new coalitions between different intellectual, aesthetic and political positions. They are subverting the critical division between 'thinking' and 'doing', emphasising the embodied character of the production of theory and the reflexive and situated approach to practice. Katie Lloyd

---

5 Maria Puig de la Bellacasa, 'Feminist knowledge politics in situated zones: A different hi/story of knowledge construction' *(womeninweb.women.it/cyberarchive/files/puig.htm)*

Thomas talks in her piece about ways of working that can escape the straitjacket of being 'for' or 'against' and produce unknown outcomes, which can exceed these oppositions. She uncovers the importance within feminist practices of shifting from 'practices of the other', to practising 'otherhow'.

Ruth Morrow's piece is about an 'otherhow' pedagogy in architecture, which allows students to learn from outside (of academia), and from others (than architects and teachers). This pedagogy for the first years emphasises the creativity of the everyday life and asks students to keep alive their memory of being *ordinary users*.

She also addresses architecture and architectural education as 'contested' domains, and proposes alterative ways to re-conceptualise them as processes of managing conflicts rather than necessarily trying always to resolve them.

A number of articles in *Altering practices* also propose an 'otherhow' approach to technology, a materialist thinking of technology from a perspective which places the body in critical continuity with the technological realm. Some of the authors, such as Niran Abbas, continue the critique of feminist science theorists such as Donna Haraway, and reconsider the relationship between the body-politics and the techno-science politics. The contributions of Francesca Hughes, Jennifer Bloomer or Sadie Plant address the interactions between matter, technology and the body through limitations and sustainability of the transformations they provoke: material, psychical, affective, social, environmental.

Ways of (be)longing

The contributors look at these transformations as *embodied genealogies*. They are interested in both *where they belong to* and *how they become*. For example, Sadie, Jennifer and Francesca point out the repression of matter and materiality within Modern architecture and theory and its 'return' in the contemporary architectural discourse. Given the traditional association between *matter and mater*, this kind of repression could obviously be related to the political repression of women's contribution to Western culture. Francesca and Sadie are reconstructing the legacies and filiations of this 'return' of 'matter' in contemporary discourses on space and architecture, convoking also the return of 'mothers' by mentioning, for example, Jennifer Bloomer's early work on 'dirt', 'formless' and 'ornament' in architecture, and that of Irigaray on 'fluidity' in philosophy. Jennifer's piece in the book shows how all belonging is rooted in a *longing*, which always acts in a realm defined by the *matter/mater* politics.

Discussing issues of identity and embodiment within technology and cyberspace, Niran draws attention in her chapter to the patriarchal tensions existing in any form of socio-cultural product, and the need to situate female subjectivity within a gendered and politicised context in the cybercultural matrix. The fusion of machine and organism becomes a progressive and transgressive site for political activity only if it involves the deconstruction of 'incompatible' frames of reference.

Ruth's pedagogy emphasises the importance of opening up the architectural education to other disciplines and processes and observes how, when internal dialogues are exposed, the reactions of others enrich the meaning and progression of these dialogues. This pedagogy prepares nomadic practitioners in architecture, who act by questioning the traditional hierarchies between client, architect and builder and by multiplying relationships and fluidities within the architectural processes.

Meike Schalk discusses 'connectionist practices' and how, by borrowing methodologies from other domains and collaborations, urban designers could act as 'curators', as creators of meaning through unusual and unexpected connections, rather than 'planners' who impose their patronising meaning on space.

The practices of collective reconstruction need maps and tools to read the new and rapidly changing spatiality of knowledge and power. They are grounded in a politics of location that maps out the different material conditions of our lives, but at the same time the way these conditions stand in relation to each other in terms of power and inequality. They are both maps and 'agencies' showing the work of signification and resignification.

In this sense, Ruth's 'pink booklet', which involves students in the sharing of the transmission of knowledge and the discourse that constructs their education, is one such map, showing how power relationships could be altered within the academic processes.

Marion von Osten's account of the exhibition *Sex & Space* introduces the work of artists, architects, researchers and theorists who investigate how space as a social, cultural and political structure produces specifically gendered identities and how the dominant construction of difference has been deeply inscribed in our modern notion of space. They identify and criticise discourses and modern legacies that support both ethnic and gender-specific exclusions within the contemporary capitalist city, realising that 'power is not just patriarchal power against women, but it is subject to constant production, reproduction and attribution processes'. In response to this, the exhibition device is used as production studio for counter-representations and non-hierarchical collaborations that deconstruct knowledge and power relationship and make visible forms of everyday work that usually go unnoticed. To construct this 'other' space, they critically use text, photography, film and the mainstream technology of communication and display (like reality TV, the internet and movie sets), emphasising with irony their materialist *détournement* and manipulation.

Francesca Hughes talks in her piece about the critical use of drawing in architecture, that for a long time has been the tool of the male representation of 'form' based on exactitude and precision, a type of representation that has always failed to measure 'matter'. This is because matter doesn't deal with precision, similarity and repetition, but emergence and difference... So what, then, will be the means of representation or rather re-presentation of matter (in space)? Katie Lloyd Thomas's chapter introduces her critical engagement with architectural drawing from a feminist position, exposing her concerns with its omissions, repressions, hidden motivations and operations. She locates this critique in her own practice of drawing, which is rooted in processes of collaboration, translation and reciprocal interpretation between 'drawing' and the 'text'. She is interested in both the materiality of drawing and that of the written language, in their gendered and embodied nature, in their interdependency and desiring temporality.

Taking on Joan Retallack's work on feminine textual approaches, Brigid McLeer shows in her chapter how these discourses are not always concerned with altering and making sense of what has been produced within the existing norms and conventions but are challenged to open up explorations into forms of *'unintelligibility'* at the frontier of the 'feminine', into what radically confronts reason and established meanings, into what is inaudible or invisible to dominant thought and ideology. The task of feminist reconstructions is to figure this unintelligibility as 'a method of invention – a process, a "rethinking", that reorders the social, political and material culture'. It is a method that acts from a feminine position understood as 'a remote position, necessarily outside and tangential, that mobilises that position as an alternative place, not separate from, but active upon, dominant sites and discourses'.

The collective presence of the authors in the book is located within a full range of practices from architecture to writing, pedagogy, activism, media, civic performance and art, mapping a spectrum of viewpoints, positions, cultures and sites located in the United Kingdom, France, Switzerland, Germany, Sweden, Senegal, Israel/Palestine and the United States. This is the 'we' of the *Altering practices*: a collective, heterogeneous and multiply situated subject, drawing on different formations with each subjective instance, in each chapter of the book. Butler talks about the 'tenuous and phantasmatic' status of the pronoun 'we' in feminist theory, always having to deny its complexity and indeterminacy and constituting itself only through the exclusion of some part of the constituency that it simultaneously seeks to represent.[6]

Yet, this volume enacts once again this performative embodiment of the collective feminist subject – always at risk when it is pronounced because already in movement, in continual transformation, de-centring and reconfiguring alliances within a shared political and poetical project. Following the feminist imperatives of 'being together' and 'reconstructing social relations in a different way', the

---

6   Judith Butler, *Gender Trouble: Feminism and the Subversion of Identity*, London: Routledge, 1990, p.137.

Altering practices draw on a politics of affinity, of alliances, of contaminations, of re-appropriation, of becoming.

'Poetics of relation' and the power to produce difference
Although in the *Alterities* conference there were many men participating, moderating and giving papers, one might note that in this book there are only female authors. This was not a deliberate decision: this omission resulted from a series of alea and transformations induced by the long editing process within the book. In order to correct it, I didn't look specifically for male contributors to play the token role. The book has consequently become a kind of 'room of our own', a piece of feminine textuality, a collective bag woven by women's voices to host their yearnings on space and transformation. And even if one could argue that not all practices included here are explicitly 'feminist', feminist strengths can be identified in the way they position themselves within their professional, social and political context. They are feminist in the way they *take place* and *alter it*.

This multiple connection between pieces resulted in a book which is not divided into thematic sections but takes its shape from a 'mise-en-relation entre des voix-femmes' in which each piece negotiates its proximities within a text like a rhizome, in which any identity extends itself towards another... Most authors in the book know each other, have previously met and collaborated; they cross-refer and comment on each other's work in their chapters, sometimes even still work together and form collectives, as is the case with the Taking Place group. The way contributions are written carries on this performative *being, thinking* and *acting together*, within a rhizomatic configuration, which conserves the multiplicity of singular positions.

Feminist textual practices are less based on interpretation than on decoding the network of connections and effects that link the text to an entire socio-symbolic system. They embed both a creative and a critical dimension. As put by Braidotti, the feminist text is materialist, is a term in a process, that is to say, a chain reaction encompassing a web of power relations.

That's why relations and connections are not always simple. Jane Rendell speaks in her piece about her long-term critical relationship with French feminism which is structural to her practice of theory and activates the conflicts between her theoretical knowledge and her political aspirations. She points to the difficult and controversial relations with power and authority, with the values of patriarchal capitalism that run right through academic research and the institutional power structures, that even feminist approaches might reproduce without acknowledging.

Liza Fior, in Chapter 4, underlined Matrix's legacy in the practice of muf but also how, in male-dominated contexts, such as the architectural profession, women's practices are automatically related to and compared with other women's practices just as women tend to be compared with other women. Still critical of the fake relationism, muf base their work on a relational logic and explain that one of the conditions of making a project successful is to build creative relationships with its clients and users. They point to the understanding of the 'personal' as 'political' and the expansion of this principle as a method of brief development. There is a materialism in the way a project can be organised and managed which influences the materiality of the resulted objects. Before being a building, the project is a *rhizome*. The figure of the *rhizome*, borrowed from Deleuze, and much discussed lately in architectural theory, belongs to the poetics of relations. For Braidotti, 'difference' is always to be considered within a relational condition.

Anne Querrien speaks in her chapter about 'difference' as 'power to produce difference'. She addresses the need to create forms of public space that welcome this reproductive power of difference: openly productive spaces that decode themselves as spaces to be continually re-appropriated and transformed...

Ilana Salama Ortar criticises the consumerist public space in Haifa, which erases difference within the social and political memory and transforms everything into commodity. Ilana's work tries to 'thwart' the effects of this erasure, by bringing back into the public realm the image of what has been forced to disappear under the ironic form of an object for sale that questions, by its strange and unexpected presence, the nature of the whole process of erasure.

'Thwarting'

'Thwarting' is, according to Jacques Rancière, the specific mode of political activity:

Political activity is that which removes a body from a location to which it was assigned, or that changes the determination given to a location; it permits the seeing of something no location was supposed to have seen, permits speech to be listened to where formerly only noise could have been heard... Politics exists when the natural suppositional logic of the ruling power... is thwarted... Politics are not made by power relationships but by relations between worlds.[7]

'Altering' could sometimes mean 'thwarting', which is also a spatial word. The *Altering practices* are political in their way of thwarting, of interrupting continuity and making room within the dominating rules in academia, professional practices, public life... They are also poetical, *poethical*, to borrow the term suggested by Brigid, in their way of making other worlds possible and suggesting other solutions to the challenge of reality than mere *adaptation*. Their ways of thwarting involve resisting homogenisation, normativity, regularisation and automatisms, not only within the profession but in everyday life itself. This political 'thwarting' is at the same time a form of 'critical openness' which, as suggested by Jane in her chapter, could leave space for other economies than those induced by the ruling power of the 'patriarchal capitalism', for new models of reciprocity, new two-way exchanges, patterns and dynamics of equivalence.

Micro-strategies, minor architectures

Helen Stratford talks in her chapter about 'micro-strategies of resistance' that are located within the realm of the 'micro'. Feminist practices are dealing with micro-strategies as localised resistances that can respond to individual oppressions and thus allow for a plurality of statements and a complexity of identities. 'Micro', for her, refers to 'the rejection of any general

---

7   Jacques Rancière, *Disagreement: Politics and Philosophy*, Minneapolis: University of Minnesota Press, 1999.

or universalising emphasis'; it indicates 'not a difference in scale but a difference in kind'. Also, she theorises the *detail* which interprets the 'micro' as less a specific scale of looking than a *field of inquiry*, 'a dispersed framework through which to re-view space and spatial relations'.

Following feminist interpretations of Foucault and Guattari, Helen shows how this logic leads to the possibility of defining diffused powers and politics at the 'molecular' level of subjectivities and comments on the 'micro-politics' developed by the feminist group Matrix and discussed by Julia Dwyer and Anne Thorne in chapter 3.

'Micro' is related to the 'minor', another mode of operating within feminist practices. Referring to Jennifer's idea of 'minor architecture', Kim Trogal brings in her chapter the example of a project for an 'open kitchen', which uses the language of cookery to form a political statement regarding the processes of building in architecture and to criticise urban development conditions in the post-industrial city of Sheffield.

Ilana's Civic Performance Art projects demonstrate how a minority point of view works always within an articulation of scales and kinds. She points directly to the Israeli-Palestinian conflict by reconstructing the model of a demolished Palestinian house in the hall of an Israeli supermarket, which was erected in its place, thus indicating the contiguity between her own 'micro-politics' and the geopolitics shaping the context of her project.

In another context, the micro-politics of urban curating adopted by Meike Schalk's team in Stockholm were the critical alternative to approach the large-scale regeneration of the harbour area.

'Politics & poetics of (dis)location'

The gendered, racialised and classed identities are 'fluid', being constructed and sustained by geographies of place, space and time and being produced 'in different ways in different places'.[8] *Altering practices* are concerned with

---

8   Geraldine Pratt and Pauline Hanson, 'Geography and the Construction of Difference', *Gender, Place and Culture* 1(1) (1994): 9.

'politics of location', a term that we borrow with Braidotti from the feminist poet Adrienne Rich.[9] A *location*, in Rich's sense, is both a geopolitical notion and a notion that can only be mediated in language and consequently be the object of imaginary relations.

That's why the politics of location are also poetics of location – they imagine the place taken by the feminist subject and at the same time the way in which this place is removed from the dominant culture. Because, as muf point out in their piece 'an invisible privilege', the location is not always a 'privilege' and could restrain the possibility of creativity which seems dependent on some sort of hidden spatial matrix of social and political circumstance… That's why this location should sometimes be understood as a dis-location, an unexpected shifting to other meanings and places. *She enters by stepping on a 'stray sod'*, Brigid says in her chapter.

The 'political' and the 'poetic' are not opposed but are chiasmatically exchanged within the *Altering practices* that talk about politics that are always poetics: forms of *poiesis*, 'ways of making', of working with space which are common to all the contributors of this book. They explore modes of practising and writing that refuse reductive binary oppositions, denounce patriarchal structures, criticise economic and political dominations and instigate new political and poetical regimes based on alliances, re-appropriations and becomings.

---

9   Rosi Braidotti, op. cit., p.21.

Chapter 2

bridge

Teresa Hoskyns, Doina Petrescu and other mixed

Frances Bradshaw
klt

TAKING PLA
AND
ALTERING I

Maria Puig de la Bellacasa
Edward Soja
Matrix
Rosi Braidotti

Julia Dwyer
helen
Helen Stratford, Katie Lloyd Thomas, Teresa Hoskyns

Judith Butler; Frances Bradshaw

Chapter 2   Teresa Hoskyns, Doina Petrescu and other mixed voices

Kelly Ives

Jane Rendell, Barbara Penner and Iain Borden

Ben Campkin

Doina

TH

TAKING PLACE AND ALTERING IT

Six years after *Alterities* and not having attended the event, Doina asked me to co-write a piece about the 'before' and 'after' of the conference.[1] During the conference I was working at Anne Thorne Architects in London: Anne Thorne, who had been a founder member of Matrix, had gone to Paris with Julia Dwyer to present a joint paper about Matrix. What I experienced at the time was the excitement that followed the conference about the possibilities of new directions for feminism in architecture.[2]

Rosi Braidotti notes a divide that existed in the 1980s between Continental and Anglo-American positions in feminist philosophy. Continental feminists such as Cixous, Kristeva and Irigaray were 'difference-inspired', especially those practising *'écriture féminine'*: Anglo-American feminists were rather in the camp of 'gender opposition', arguing that 'sexual difference' was essentialist, and sought a 'beyond gender subjectivity' that overcame sexual dualism.[3]

From a British practice point of view, *Alterities* was an event that enabled some UK feminists to come back from Paris with the tools to talk about an architecture of the feminist/feminine without being accused of essentialism. Although the theory was Continental, architectural practice in the UK was more sympathetic to feminist discourse, and groups like WAFER (Women Architects for Equal Representation) in London were meeting monthly at that time.[4] Jane Rendell was bringing together Continental feminist and architectural theories in her doctoral research and through her teaching at Nottingham

---

1 This invitation follows a forum discussion hosted on *www.inplaceofthepage.co.uk* involving a few of the founder members of the feminist group 'Taking Place'. Taking Place emerged just after Alterities, its activity mapping somehow this period that separates the original event from the current publication. The forum was meant to question this relationship. Comments selected from this exchange are quoted in several notes. The forum started with the following questions: What from 'Alterities' has contributed to the emergence of 'Taking Place'? How 'Taking Place' could redefine now 'Alterities?' (posted 01.30.04 22:51:55 by Doina)

2 I find the questions difficult to think through and want to come back to them in another reply, but am concerned that they are so much less answerable for those who were not at Alterities - particularly Teresa, since she has been a founder of Taking Place. Maybe we need to map out the other constellations that have been part of it - such as WAFER, which is where Teresa and I met (because I bumped into Julia who started WAFER many years ago) and talked of something bigger, Jane's seminars at Nottingham and so on (there must be many more!)... This is not to say that Taking Place did not grow out of connections at Alterities, only to say that there is not one line. (posted 02.09.04 16:35:54 by klt)

3 See Rosi Braidotti, *Nomadic Subjects*, New York: Columbia University Press, 1994, p.140 and *passim*.

University (where Katie Lloyd Thomas and Helen Stratford were her students). The Nottingham feminist group held a specific panel in *Alterities*.[5] Jane was also co-editing a book, *Gender, Space, Architecture*, which introduced different feminist orientations in relation to space and architecture, including continental feminist theory.[6]

This chapter describes only one of the many possible narratives of the story of feminist practice in architecture, a narrative connecting British and French feminist positions at a particular moment in time, shaped by specific circumstances and recalled in personal memories and interpretations.[7] What is feminist can be described as 'a movement of struggles to change values'.[8] Feminism can be seen as continually shifting away from those values that become appropriated and appropriating values from other struggles. In both theory and practice, feminism has a continuous and dialectical relationship with non-feminist struggles and mainstream culture. We

4 WAFER's story started with a network of women architects in London who met intermittently from the early 1980s till around 2001 in London. Political discussions and actions in the late 1970s and early 1980s resulted in the creation of number of feminist organisations operating within the field of the built environment (Matrix, Women's Design Service (WDS), Mitra, Women and Manual Trades (WAMT), London Women and Planning, Women in Construction etc.) which provided a supportive background for networking amongst feminist architects and designers. As FAN (Feminist Architects Network), the network was responsible for two day events (Women in Architectural Education in 1983, Feminism vs Professionalism in 1984) and a weekend event which attracted over 60 people ('Women's Realm', Jan. 31/Feb. 1 1987), as well as an evening course on the building industry for women builders in collaboration with tradeswomen from WAMT. After a gap from 1987 until the mid-1990s, the network revived as WAFER in response to the Labour Party's election manifesto on architecture released before the 1997 election (when the first Labour government was elected in the UK since 1979), and met the Labour Party Women MPs Group to discuss women and architecture (specifically the MPs' influence on the designs for new offices for MPs just being completed). WAFER met regularly as a critical forum to discuss issues including architecture and health and architectural photography and representation. Architects, artists and academics presented individual architectural, art and writing projects to a critical and supportive group. The connections made within WAFER were behind a series of other initiatives by smaller groups, both academic and professional. (Excerpts from Julia Dwyer's email exchange with Doina Petrescu, 30.11.2005.) See also Julia Dwyer and Anne Thorne's chapter in this volume.

5 For a personal account of the problematic relationships between the French and British feminist legacies and how they have been reflected in the organisation of this event, see also Jane Rendell's chapter in this book.

6 Jane Rendell, Barbara Penner and Iain Borden, *Gender, Space, Architecture: An Interdisciplinary Introduction*, London: Routledge, 2000.

7 By writing this story six years after the *Alterities* conference, we have noticed that things have certainly changed since, but not so much. We have learned something about the temporality of some feminist projects which operate with slow moves, facing constraints and taking the time needed, not following the imperative of fashion and 'efficiency'.

8 R. Braidotti, op. cit., p.142.

are therefore going to tell the story of 'our' feminist practice as a series of shifts rather than trying to make definitions, using one thread but knowing that many other stories could be told.[9]

Before *Alterities*

The first shift in thinking is that from Matrix to *Alterities*. Matrix started in 1980 after the split of the Feminist Design Collective,[10] formed in 1978. It is one of the historic feminist groups that were invited to participate in *Alterities* and is described in Chapter 3 by Julia Dwyer and Anne Thorne.

Theoretically, Matrix can be seen as operating within the Anglo-American feminist school. It was concerned with 'making space' by women who designed in a man's world and argued that men and women were *equal* as architects. 'Sexual difference' was something to be argued against, as it placed women within the private realm.[11] It was concerned that space should *include* women by directly involving them in all stages of the design. Matrix found it could make spaces that were more suited to women's lives, but it found it difficult to define an aesthetic difference between architectural designs made by women and those made by men, due to the influence of architectural training.[12] At that time much architectural training, like political thinking, could be seen to be about universality and rationality and modern architecture reflected this; architectural students produced similar work, regardless of context or background. Bradshaw wrote that in education, women students' work would often be devalued as 'emotional or confused'.[13]

One of the problems that has been continually addressed within feminist discourse is whether a 'truly' women's art can exist. Can women make art that is completely free from

---

9 I agree with Katie about broadening the coordinates of this alterities/taking place/etc. network. And for me tracing the routes of our various ways in to the discourses and the settings in which those discourses are 'taking place' at the moment, is one way of embracing those histories and trajectories. (posted 2.6.2004 10:00:58pm by bridge)

10 Feminist Design Collective was set up to develop a feminist approach to architecture through discussion and architectural work. See Frances Bradshaw, 'Working with Women', in Jane Rendell, Barbara Penner and Iain Borden (eds), *Gender, Space, Architecture: An Interdisciplinary Introduction*, London: Routledge, 2000. See also Julia Dwyer and Anne Thorne's chapter in this book.

11 See introduction to Matrix, *Making Space*, London: Pluto Press, 1984.

12 Ibid., p.282.

13 Ibid.

patriarchal influences?[14] In fact, it is only through practice and critical reflection on practice that feminist art or architecture develops.

It was in practice that Matrix showed signs of what was to follow. Matrix were critical of much modern architecture: in the Jagonari Asian Women's Centre in Whitechapel, Matrix worked with its female clients/users on decorative components for the outside of the building.

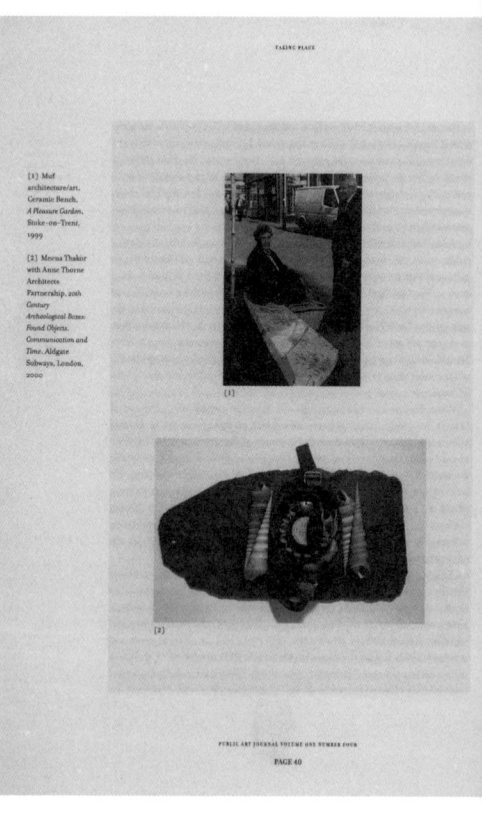

Public Art Journal, volume 1, number 4, October 2000

'Bangla Town' gates, a project by Anne Thorne and Meena Thackor (Credit: Ben Campkin)

---

14 See Kelly Ives, *Cixous, Irigaray and Kristeva: The Jouissance of French Feminism*, Worcestershire: Crescent Moon Publishing, 1996.

At the time of *Alterities*, I was writing an article for the *Public Art Journal* that asked this very question. I was interested in the way that feminist practices such as muf architecture/art, Anne Thorne Architects and Dolores Hayden were collaborating with public artists on projects using long participatory processes. They used public art to almost decorate architecture in order to represent those whom they saw as aesthetically marginalised. What was emerging was a diverse and differentiated aesthetic. Could this architecture be described as 'feminist' or 'feminine'? Projects like muf's 'A Pleasure Garden of the Utilities', a competition organised by the City of Stoke-on-Trent for the first public art installation within the City Centre Gateways Project, celebrated the traditional craft industries of the city by designing two benches fabricated from the ceramic that Stoke is famous for, patterned with a distinctive blue and white floral design. The benches were made at the local ceramics company, Armitage Shanks, where the artists worked closely with the production and design divisions. Two local companies then applied surface transfers.

The strength of this project came from the involvement of local networks. As the pieces passed through the factory, people became involved in the artwork, creating a shared ownership, although muf still retained authorship. Like the plates in people's kitchens, the pieces were very obviously from Stoke-on-Trent, emphasising the connection between the work and a powerful sense of place associated with the town.[15]

The Aldgate Subways project by Anne Thorne is typical of what I would describe as 'identity'-based architecture. Like Dolores Hayden, she is concerned with the male architectural reading of the city. Aldgate Subways, in the newly branded 'Cityside' area of East London, was created to address some of the social and economic disparities that exist between Tower Hamlets and the City. The uniformity of the subway system with 28 exits not only meant that it was very difficult to find the right exit, but that the subway was a place feared by women.

15 Teresa Hoskyns, 'Taking Place', *Public Art Journal*, 1(4), October 2000, pp.37-43

Thorne's methods included working with three artists to organise the area into colour zones, closing some subways, and providing new crossings. The artists' work involved lining the walls with enamel panels, installing canopies above the exits, and building a glass archive of artefacts collected from local businesses and residents to be placed next to the panels. This patchwork of ideas also included Thorne's own designs for railings. Thorne was thus transforming a clean-cut 1960s subway system by collaging it with contributions from the local community to recreate difference within the perception of place. She used decoration to represent those she saw as 'marginalised'.

Ben Campkin argues that the work becomes problematic when artwork is used to claim space for particular sections of the community. His example is Thorne's 'Bangla Town' gates at the entrance of Brick Lane, 'an area that has been able to be adapted and adopted by so many different groups of immigrants with such success' and has recently seen a process of gentrification with an influx of higher income people moving to the area.[16]

Alterities theorised a shift from feminist practices of identity to practices of difference. The conference discussed the shift from politics to poetics. In this respect, *'écriture féminine'* was invoked as an attempt to recognise specificity within female writing. Irigary, Cixous and Kristeva use different voices in their writing, mixing together feminist, political, poetic and confessional modes. The conference was followed almost directly in our practice by a second shift that interprets difference within spatial politics. This is explicit in the experience of the architectural feminist practice and research group *Taking Place*.[17]

---

[16] Ben Campkin, 'On Ghettos: urban boundaries and cultural identities', Lecture at University College London, 24 November 2005.

[17] There was a move from politics to poetics, from 'feminist' to 'feminine' and a new aesthetic was emerging. I met Katie at WAFER and at that meeting the discussion was about whether women have equal opportunities to men in architectural offices. Some of us didn't want to discuss that any more. Alterities has been used to critique WAFER and from that critique emerged Taking Place. The starting point was that something has been already achieved: the question was no longer about equality but difference. Also something new was emerging and needed placing.(posted 2.8.2004 7:09:51pm by TH).

Taking *Alterities* place

Taking Place is a spatial practice of architects artists and writers. It was founded in 2000 by ten women who began by wanting to organise a feminist architectural conference. This became a conversation about architecture, art and politics over many months of meetings.[18]

Taking Place suggests a fundamental shift from previous feminist architectural work, because we were writing from an architectural perspective that can be read from other perspectives as well, including art or performance. The group has moved away from the aim of representation (as seen in many public art projects), and from an understanding of the architectural space as space formed by objects in relation to one another, to a more political understanding of the space-time relationship.[19] The notion of participation has shifted from a participation in the production of objects to a spatial participation. The aims are no longer to be 'included' or 'represented' but to participate directly from a differential position. This is the legacy of *Alterities*: 'difference' became a tool for 'taking'.

The development of this position was influenced, most importantly, by Luce Irigaray and her work on *The Ethics of Sexual Difference*, translated into English in 1993, Rosi Braidotti's on 'nomadology', and Judith Butler's work on 'performativity'. In her seminal book, *Gender Troubles*, published in 1990, Butler argues, appropriating Foucault, that 'gender identity' is constructed along 'culturally intelligible grids of an idealised and compulsory heterosexuality'; she suggests that reality is fabricated within some sort of interior essence. This interiority defined by signifying absence is 'performative'. It can only be suggested and revealed through playful enactment.[20]

In *Nomadic Subjects*, Braidotti appropriates Deleuze and Guattari's concept of nomadism, to provide an alternative to the essentialism versus sexual difference debate:

---

[18] The founder members of *Taking Place* included Jos Boys, Miche Fabre-Lewin, Julia Dwyer, Teresa Hoskyns, Katie Lloyd Thomas, Brigid McLeer, Angie Pascoe, Doina Petrescu, Jane Rendell and Helen Stratford.

[19] Following Lefebvre's *The Production of Space* and Harvey's *Social Justice and the City*, Soja notes that 'space and the political organisation of space express social relationships but also react back upon them'. Edward Soja, *Postmodern Geographies*, London: Verso, 1989, p.76.

[20] Judith Butler, *Gender Trouble: Feminism and the Subversion of Identity*, London: Routledge, 1990, p.137.

The nomadic subject I am proposing is a figuration that emphasises the need for action both at the level of identity, of subjectivity, and of differences among women. These different requirements correspond to different moments... different locations in space... different practices. The multiplicity is contained in a multilayered temporal sequence whereby discontinuities even contradictions can find a place.[21]

> Taking further Irigaray's ideas, she locates the feminist political project within the idea of a *practised difference*. She says also that 'difference to be operative has to be acted ON and acted OUT collectively'.[22] She emphasises 'on' and 'out' in her phrase, on the necessary performativity and transmission of the experience of difference, but we would also place a stress on the word 'COLLECTIVELY' which is also very important for us. The feminist political project of a practised, performative difference is necessarily a collective project. 'Difference' is a relational concept. It cannot exist outside and has always to be established within a relational condition. One can only be different in relation to others.[23]
>
> This is how we have performed difference with the *Taking Place* group, always perceived as a collective, relational practice, which provided not only a 'common place', but also a multiplicity of places taken simultaneously by our different subjectivities.[24]
>
> The 'nomadic subjectivity' at work in *Taking Place*, not only has freed us as architects by allowing us to move from site to site, but has allowed us to move across hierarchies and disciplines, in our case, into writing, art and performance.
>
> By expanding the notion of space into writing, performance and art, *Taking Place* wanted to create an extended space

---

21 R. Braidotti, op. cit.
22 Ibid., p.145.
23 Ibid., p.32.
24 What are our politics-poetics?... Maybe we can't simply define them because they are not just 'ones' but different, they are not just consensual but also conflictual, heterogeneous, altered... that's why the initial taking place group has changed, it seamed that we had different interests that we cannot accommodate... that we have chosen to accommodate our identity as a group, to make our group identity follow the differences by altering our number and configuration according to every instance of the 'taking place'. This made us aware of not only our 'common place' and shared identity but also of our (radical) difference, of the different places we take simultaneously when we act as a group. (posted 03.02.04 21:32:38 by Doina)

within the public realm and a redefined discipline of architecture where feminist discourse can be explored.[25] The feminist notions of 'location' and 'situatedness', as suggested by the syntagm that names the group, refers to both physical, cultural, professional and virtual space rather than the more traditional notion of architectural space which refers only to a physical site. But, as stated by Maria Puig de la Bellacasa – a feminist who takes further Haraway and Harding's lines:

situatedness is also an *act*, a gesture, a political gesture. It doesn't function as an accusative external 'moralistic theory' but as an embodied gesture of affirmation: we situate ourselves. As such, it is better propagated by contagion, not as a normative framework.[26]

Being aware that knowledge is always produced from *somewhere, Taking Place* has deliberately chosen to situate itself as a group acting between academia and other practices from outside academia. We have politically chosen to practise difference ON and OUT, and propagate it by 'contagion'. We have tried to find out how difference could be enacted in space, in the same way as Foucault demonstrated how power could be directly expressed through spatial relationships. We have started to think of 'difference' as 'power'.

Before us, Matrix concentrated on how women architects, planners and clients might take part in the making of built space.[27] By asking 'How do we use architectural skills to further the liberation of women?', it found that it was possible to build buildings that respond better to women's needs but found it difficult to define what a radically different architecture might be.[28] By considering space as more fluid, both physically and socially, *Taking Place* has discovered that it is possible to produce another quality of physical public space

[25] So I guess for me, 'taking place' slowed down 'alterities', stretched out the intensity of that conference, and became a much more performative version of the 'staging' inherent in the conference. (posted 02.06.04 22:00:58 by bridge)
[26] See Maria Puig de la Bellacasa, 'Feminist knowledge politics in situated zones: A different hi/story of knowledge construction'. (*womeninweb.women.it/cyberarchive/files/puig.htm*)
[27] Helen Stratford, Katie Lloyd Thomas, Teresa Hoskyns, 'Taking Place', in *Scroope, Cambridge Architecture Journal*, 14 (2002): 64-8.
[28] See Frances Bradshaw, 'Working with Women'.

through intervening in the power relations expressed through space. Place can simply be 'taken'.[29]

Taking Place has now organised five events, each one further developing this feminist space, the practice of participating in the event and challenging preconceived theories of what the space might be.

Taking Place 1 was an event that happened over two weekends in a caretaker's house at the University of East London in the summer of 2001. Following many discussions about organising a conference and inviting 'important' feminists, it was decided instead to start by presenting our own work to each other. This process produced a rich diversity of work ranging from confessions to debates, city walks and film and slide shows, most taking place around a table with food, in a cosy and interesting atmosphere. *Taking Place 1* produced the material for a larger event, but asked how other people could be involved in this feminine, cosy, domestic yet academic space, which was our chosen mode to make the shift from 'private' to 'public'.

Taking Place 2 was organised in November 2001 at the University of North London School of Architecture. It was introduced in place of the regular Thursday evening college lecture, followed by the main, one-day Saturday event. Each of us invited ten people known to be involved in feminist thinking to this day. The college was transformed: the front window became a kitchen, lectures and performances took place on staircases, in the WCs, outside in the courtyard, in the pigeonholes of staff members. The main lecture space became a cabaret area and cooking aromas filled the air.[30] Breaking

---

[29] I have experienced both Taking Place and Alterities as 'places' through a kind of belonging to them; perhaps Alterities was, for me, a place that I could enter, while as Teresa suggests, Taking Place has been one that is taken. (posted 02.09.04 16:35:54 by klt)

[30] Part of the joy of Alterities was in the discussions that took place in cafes, in the metro, in the hotels around it - in the way it exceeded the centre of the planned event, and Taking Place is also a part of that excess and momentum. On the other hand, just as Teresa describes a transformation from WAFER to Taking Place, there is a transformation from ideas and discussions that occurred at Alterities into practices - that Taking Place 2 no longer had a conference format, its structure as an event (or performance) attempted to embody those ideas. Modes of thinking perhaps moved into modes of doing. And if we reacted against the discussion of equality that was being had at WAFER we also took with us some of WAFER's way of proceeding - eating, socialising as well as the central discussions, and the format of making short presentations (at least during Taking Place 1). (posted 2.9.2004 16:35:54 by klt)

down boundaries and forming a new space were slightly uncomfortable for participants and the main criticism of the day came from women wanting to participate more.

Taking Place 2, 'Hot Water' discussion

Taking Place 3 was held at the University of Sheffield in November 2002. It formed part of a three-day feminist 'whole school' event for students and staff in the School of Architecture. As some tutors feared that the (mostly male) students would be hostile to the event, it meant that it was open to all. Instead of a series of performances, it was a series of workshops. Instead of the Taking Place team appropriating the space, students did. Starting with a debate asking the question, 'Do women experience space differently to men?', feminism was openly discussed from the start and positions evolved throughout the day.[31] Students were asked to produce spaces, writing, objects, arguments and interventions. 'Contaminations' was a series of spatial interventions. 'Talking Places' was a series of discussions combined with spatial transformations. 'Writing Instead of Speaking' gave students the

---

31 For me 'taking place' picked up on the exciting notion of a constantly evolving action, as we wrote for the Berlin proposal 'what began as a shared and indefinable loss of place has become a process; a practice. We have, no fixed place and no identity apart from our encounters. This unfolding process that we have initiated forms our practice.' Like Taking Place 3, this proposal, and the e-mail forum began for Iceland, aimed to engage a wider public audience to create its space: as the proposal continues:
'a means by which to have place and a desire for others to join in this critically assertive and productive place'. I think this echoes a lot with what Brigid was saying about opening the space out... (posted 2.23.2004 21:29:55 by helen)

Taking Place 3, 'the very act of speaking' walking tour and 'not teaching' performance
Below: *Taking Place 3* timetable

taking place 3: timetable

when? **what?** **where?**

wednesday 06 november 2002
2.00 the event begins — double height studio floor 16/17
form teams, collect workshop packs

thursday 07 november 2002
11.00 *talking places 1* the great taking place debate — crit space 15A and B
in this Oxford Cambridge style formal debate, hear 2 teams present the case for and against a hotly contentious motion about space and gender, followed by audience questions and a vote on the motion. Chairs: Doina Petrescu, Jeremy Till

12.00 'speaking in place of listening' — floor 19 lecture hall
Writer Jane Rendell will read her piece 'speaking in place of listening' where she once listened to lectures herself.

1.00 'the very act of speaking' walking tour — starts floor 19 lecture hall
Join Jane Rendell and students on a performance walking tour around the school to discover how this workshop has explored the relationship of identity, voice and place and produced site specific interventions using 'voice' and/or 'words'.

2.00 grab a sandwich

2.30 'contaminations' walking tour — meet outside y1 studio floor 13
Discover the spatial dislocations and contaminations which have been created by this workshop to intervene within the existing power structures of the school on a walking tour with architect Helen Stratford and students.

3.00 'in place of the page' presentations — venue to be announced
See the workshop presentations of a 'kiosk' for stories about place, in which processes of recording, storing and reading stories about place, could happen and join in the discussion with artist Brigid McLeer and students.

3.45 'not teaching' performance — venue to be announced
Watch a performance show with students, teachers and non teachers about 'What means not teaching for you?' in a surprise venue proposed and designed by students. This workshop is proposed by Doina Petrescu in collaboration.

4.45 *talking places 2* 'speaker and audience' talks — MA lecture hall floor 15
Find out from Teresa Hoskyns and Katie Lloyd Thomas about the ideas behind the 'talking places' events through short talks and some films from the West Euston Project, in a lecture space designed by students with some spatial surprises.

5.45 *talking places 3* the last word – everybody! — double height studio floor 16/17
Please join all visitors, staff and students for 'last words' – an unplanned event for the whole school. You will have the opportunity to form groups based on a shared identity (real or imagined) and have your views heard or seen by all.

6.30 evening feast – all are invited — first year studio floor 13
Food constructions> Throughout the day Kim Trogal, Doina Petrescu, Leo Care and students have been working with your food contributions to transform the first year studio into a space for sharing food and enjoying eating it together.

8.30 social taking place – a relaxing drink or two — sola bar in city center
You are invited to a SUAS party organized for the whole school to follow the day's events.

*

all day Pater Noster Phobia installation — view from inside the paternoster
by Petra Marguc, curated by Helen Stratford
Scrub Gardens — in unlikely spaces
Prue Chiles, Caty Dee, Simone Abram, Stephen Walker
Food constructions (deposit your own contribution too) — in and around first year studio

friday 08 november 2002
10.00 deconstruct, coffee and records — first year studio floor 13
The food construction and docu-commentary group will be tidying, drinking coffee and reviewing the event

*

all days Docu-Commentary group will be filming, recording and interviewing to get to the inside story of taking place 3. — everywhere
workshop facilitated by ruth morrow

opportunity to perform a piece of writing. 'Food Constructions' workshop ended in a party in the first year studio.

The whole series of workshop resulted in an open participative event in a school of architecture which performed, even if only for a day, a critique of the academic institution and the architectural education, both in terms of programme and space.

Taking Place 4, forum discussion

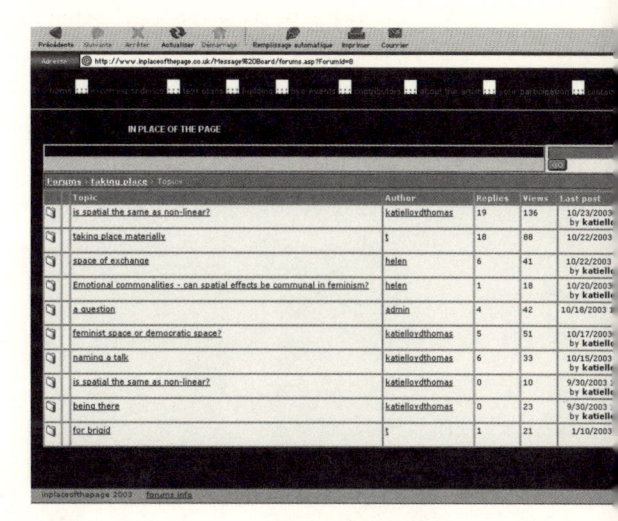

Taking Place 4 developed from an invitation to give a talk at 'Becoming Space: Ideology, Invention & Immanence in Human Surroundings', an international conference on 'Event Space' at the Living Art Museum, Reykjavik, Iceland, in October 2003. We set up a web-forum to enable us to discuss the talk without the need to be physically present and to increase possibilities for intermittent and extended participation. This recorded the extended discursive processes that characterised our meetings, which exceeded the public events they produced. Katie, Helen and Brigid talked about this place and exhibited the entire contents of the forum. The separateness and distance in time and place, as well as the recording technique inherent in the process, allowed us to articulate thoughts and ideas in a different manner. As Helen puts it in her posting:

Talking Places was an exchange that is slowed down and stretched out... strangely rigorous and focusing... also interestingly disconnected. And while the language of the posting attempted to exert hierarchies, rating of 'members' as 'senior', 'junior' and 'new', the excess of this space enabled a more intense sharing of concerns and ideas, and gave us time and space to go further than we often had the chance to, when we meet face to face.[32]

As a public space that can be freely produced, *Taking Place* became a forum where it was possible to invent a new practice and to allow each of us to explore projects that we were working on. The forum infrastructure has been used on other occasions, for example, to host the discussion about the relationship between *Taking Place* and *Alterities*.

Taking Place 5, working meeting in Julia's kitchen

Taking Place 5 took place in March 2005, when we were invited to intervene within the framework of the symposium 'Technologies of Place' held at Akademie Schloss Solitude in Stuttgart, Germany. The intervention took the form of a circular walk. Made up of a series of presentations, the walk examined 'technologies of power', technology as both product and productive of shifting socio-economic and political

---

32 See forum 'Taking Places', *www.inplaceofthepage.co.uk*

Taking place and altering it

Taking Place 5, maps and sweets

networks or assemblages. Part interior, part exterior, the presentations responded to the space of the Akademie and the conference to explore intersections of technology, power relations and spatial practices, in order to add necessary critical and political dimensions to these concepts. Areas addressed included: the critical potentialities of new media/technology in the need for new fields of action in public space, the idea of the virtual within the everyday, the temporality of a dialogic space of an Internet forum, and the restructuring of individual and collective subjectivities within new technological contexts. In this context, new media and technology were less the distinct object of discussion than part of an assemblage of practices, incremental methods or enmeshed ways of working.[33]

Altering *Taking Place*

Taking Place has been a 'practised theory' shaped from inside academia. How has this informed architectural practice? First, we have expanded the understanding of the architectural practice. This dropping out from conventional architecture has been part of a strategic feminist political struggle of liberating the architectural practice from its restrictive norms and regulations (such as those defined by professional bodies and centralised institutions like the RIBA and the ARB in Britain).

[33] 'Technologies of Place' Helen Stratford's introduction to the symposium.

Another concern has been the coexistence of individual approaches with the possibility of collective practices and the multiplicity of forms of collaboration that a feminist practice can take in architecture. We also felt the need to go outside the feminist 'club' (and its academic location) and expose our approach to other practices and other social contexts.

In other words, how to 'take place' both inside and outside of the profession and what to do with this 'place'? How to open a place once 'taken' and how to define feminism in architecture as an inclusive, participatory spatial practice, which doesn't belong necessarily only to architects?

This re-thinking of architectural practice was at work in the 'Open Spaces for All' project in Regents Park Estate, Euston, London.[34] We discovered through this project that beneath the homogenous architecture of a council estate lies a multiplicity of identities, tensions and complex relations that are played out by, among others, young people through their occupation of the estate's open spaces. The project addressed these sensitive relations while acknowledging that there were no easy architectural solutions. No building was commissioned, and this absence of object changed our approach, enabling young people to play out inner desires and fantasies through taking part in a process of occupying, performing and creating new spaces by using digital imagery and film making. Through this experience, a different understanding of architecture started to emerge.[35] Four major projects came out of the five years' work at West Euston, each taking the same format, which involved young people becoming photographers and film makers and going on cultural outings, such as films, art exhibitions and walks. A residential exchange to Lille in France was organised to enable encounters with other people and places, and to follow the progress of a new garden designed by Yvonne Dean, John Medhurst and Sue Ridge for the Mosaic Garden in the Parc de la Deule, near Lille.

---

[34] 'Open Spaces for All' team included Yvonne Dean, Teresa Hoskyns and Delroy Beaton. The project took place in 2000-5. Funded by West Euston Partnership.

[35] One of the boys said to me 'Teresa, I've been doing "architecture" again' after attending a GLAS workshop run by the Architecture Foundation. This involved them running around Kings Cross with mobile phones and phoning the information back to the GLAS team who were writing it as headline news on newspaper stands outside the station.

In the first project, groups of young people made computer manipulations and collaged their own photographs. These altered images were projected onto a tower block in a contested territory, the main square of Cumberland Market.

'Our World', Cumberland Tower
(Credit: Sue Ridge)

The images showed the diverse surroundings and how space is contested in Regents Park Estate. For one evening, the event appeared to reclaim this space for all. The second project consisted of four films, each of them mapping out the territories that self-constituted groups of young women and young men inhabited. We found that the films gave the teenagers a public voice and a method of expression, particularly a group of Bengali girls, who had previously appeared to be absent from the open spaces of the estate. In the third project, the teenagers made their own designs into panorama collages of fantasy transformations of Cumberland Market, including images from Lille, roller-coasters and chocolate walls. The panoramas were made into banners and exhibited in London and Lille. Following this, the final project was to design a little building, to be a youth shelter. The teenagers and the architects built design models in a workshop that were later exhibited outside, and residents of the estate were invited to vote for their favourite, selecting Ajir Hussein's design.

***

Collage of Cumberland Market by young women and Sue Ridge, 2004

Being particularly interested in forms of collaboration and participatory spatial practice, Teresa and I have recently carried out a project, which overlapped the different methods and contexts that we had both experienced previously. We worked with teenagers in La Chapelle – a neighbourhood in the north of Paris – who were also users of the ECObox garden, a project that I have co-initiated in this area with other architects, students, sociologists, artists and residents as a form of collective practice.[36]

This led to our collaboration in a film involving the young users of ECObox, and mixed our projects developed in separate contexts. We worked with a self-constituted team of 10 to 11-year-old girls, assisting them in producing their own films. We were interested to learn how these girls 'take place' in the city: what their 'vision' is and the role they want to 'play' and how they express themselves politically and interpret urban space.

They chose the topic and the title of 'Impossible Mission', drawing on the imaginary of gamespys and the contemporary series of spy thrillers reviving the 1970s, a period never experienced by them but inherited through various narratives of the modernist ideal of 'saving the world'. This topic also directly addressed issues of escaping control and tactical surveillance.

---

36 *atelier d'architecture autogérée/studio of self-managed architecture (aaa)* is a non-profit association and an interdisciplinary network founded in Paris in 2001 which conducts research into participatory urban actions and aims for the re-appropriation and reinvention of public space through everyday life activities (gardening, cooking, playing, reading, producing, debating, walking, etc.).

The starting point was the realisation of a temporary garden made out of recycled materials on one of the derelict sites in the La Chapelle area, in the north of Paris. This garden, called ECObox, has been progressively extended into a platform for urban criticism and creativity curated by the *aaa*, residents and external collaborators, catalysing activities at a local and translocal level. *(www.urbantactics.org)*

For more about ECObox, see also Doina Petrescu, 'Losing Control, Keeping Desire', in P. Blundell Jones, D. Petrescu and J. Till (eds), *Architecture and Participation*, London: Spon, 2005.

The girls' team wrote the storyboard and chose the areas where they wanted to shoot different scenes. The fiction film acted as an urban tool to allow them to express their choice of representation within the city and their implicit political position, to legitimise their misbehaviour in public space and their disapproval of what they have read and understood as city spatial policies. To literally *take place* within a transformational practice.

They also chose to renegotiate gender identity, to fight, to run from place to place, to parody the male winner position within an imaginary battle aimed at saving the 'ECObox garden' faced with eviction (for real) by the city authorities. They have in a way directly addressed their condition of inhabitants of a contemporary city and have expressed the ironic position of a generation of young females for whom the city is a field

ECObox garden, La Chapelle, Paris 2004
(Credit: atelier d'architecture autogérée)

of performed identity, a space to be curious about, to be spied on, desired, fought for and claimed back. The real 'impossible mission' of the film was the imperative to invent another access to the city: one which is more creative and unpredictable, more democratic and allowing everybody to play the role of her choice, to win a battle, to take the desired place within the system of power relationships, to simply *take place...*

'Mission Impossible' 2005

One of the questions of any contemporary feminist practice is 'Who are the other feminist practices today?' We think that the contemporary feminist practices in architecture no longer constitute the privileged territory of feminist academia and the profession in developed countries. This territory is a shared territory. By defining 'taking place' as a relationship between women and power, we recognise all other movements that are reclaiming participation in decision-making and spatial transformation. Our projects have to learn from the practices of the peripheral actors of globalisation processes, such as women's organisations in the cities of the global South or peripheral actors of the urban processes with the cities of the North: immigrants, inhabitants of council estates, teenagers, children, etc.[37] Practising with others alters the way we conceive and practise architecture and frees the transformative power of *alterity* in space.

---

[37] Our more recent practice deals specifically with these issues. We can briefly mention here Teresa's involvement in the Social Forum in London. (*www.londonsocialforum.org.uk*) and Doina's practice with *aaa* as well as her collaboration with REFDAF (Réseau des Femmes pour le Développement Durable en Afrique) for the *Cité des Femmes* in Dakar. (See also D. Petrescu, 'Life Matters Making Place', in Katie Lloyd Thomas, *Material Matters*, London: Routledge, 2006.)

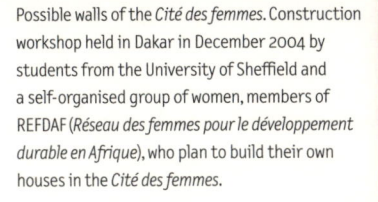

Possible walls of the *Cité des femmes*. Construction workshop held in Dakar in December 2004 by students from the University of Sheffield and a self-organised group of women, members of REFDAF (*Réseau des femmes pour le développement durable en Afrique*), who plan to build their own houses in the *Cité des femmes*.

We have learned within and beyond Irigaray's legacy how difference can become something else; how it can be used to take a different position, a different role than the one assigned to us by society; how difference is 'power enacted in space'. From the perspective of our feminist practice in architecture, practising difference has evolved into practising differently... Taking place *differently*.

Chapter 3

Julia Dwyer and Anne Thorne

Lynne Walker

Fran Bradshaw, Sue Francis, Julia Dwyer, Janie Grote, Anne Thorne, Suzy Nelson, Elizabeth Adams

Matrix

EVALUATI
MATRIX:
notes from inside the collectiv

Chapter 3 — Julia Dwyer and Anne Thorne

Sue Francis

Matrix was a book group and a design practice which was an important part of the feminist interpretation of architecture and the built environment in Britain during the 1980s. Twenty-five women collaborated over fourteen years in a co-operatively structured practice to explore the ways in which the theory and practice of architecture could respond to and affect social relations. Matrix book group produced *Making Space*, the first British collection of writings on women and the built environment, in 1984. Thirty-five women were centrally involved in Matrix, with the same number again associated with its history, and it was at the heart of a network of feminist architects, builders and academics.

Although the history of the formation of Matrix was described in 1981[1] and more fully in 1984,[2] there was little further documentation apart from the booklets and leaflets it produced in the course of its work. This chapter was developed in consultation with a small group of women who were also associated with Matrix: it has been written so as to expose past theories and practices to wider contemporary debate and to find the connections between them and our current work.

In the late 1970s, women architects within the New Architecture Movement[3] began to meet independently, and, among other actions, organised a campaign to stop sexist advertising in the building press. This practical campaign was undertaken with a confident trust in the effectiveness of political action, as NAM members were likely to have been part of or were aware of other single-issue campaigns during the late 1970s which had been successful. Many of the women at NAM meetings would have described themselves as social-ists, some as feminists, others as both. Socialist feminists considered class relations in their feminist analysis: one thing could not exist without the other.

A number of project-based groups of women architects and academics emerged, and their membership often overlapped. One group was commissioned to prepare an exhibition on

1 Sue Francis, 'Matrix', in 'Making Room: Women and Architecture', special issue of *Heresies*, 11, 1981, p.17.
2 Matrix, *Making Space: Women and the Man-Made Environment*, London: Pluto Press, 1984.
3 New Architecture Movement is a radical (socialist) grouping of architects questioning traditional architectural roles and campaigning for the unionising of architects' offices.

women and housing, 'Home Truths', which toured Britain during the 1980s. The Feminist Design Collective, a group of about twenty women was started in 1978. Its title was consciously assembled: the use of the word 'feminist' was contentious; no architectural practice in Britain had previously stated their political position so overtly. The use of 'design collective', rather than 'architectural practice', indicated the group's intention to value non-architects as highly as architects and was influenced by contemporary critiques of professionalism and of architects' professional institutions by groups such as NAM.

The Feminist Design Collective came together to work on two projects. The first of these was Stockwell Health Centre,[4] a project for a community group which was questioning the way in which the National Health Service was providing health care for the community in inner South London. Although the Health Centre was unbuilt, it gave the Collective its first experience of the dynamics of working *with* a consensus-based client group, and of the power which drawings had to represent aspirations and mislead.

Lambeth Women's Workshop was founded by Women's Aid to train women in carpentry and joinery skills, so that they could access relatively well-paid jobs in the building industry. Women's Aid was principally involved with developing refuges for women suffering from domestic violence, but was concerned with the job prospects for women after they left the refuge. An industrial building in South London was found with the Collective's help and converted by women builders. Aware of the unequal class position of builders in the production of building, and of the mutual disrespect which sometimes resulted from this, the builders argued that the building process should be far more collaborative and that methods should be adapted to make this possible, including addressing the language used by the professionally trained architects, which, they suggested, distanced them from both clients and builders.

Feminist architects were drawn to building: they wished to dissolve the class differences between themselves and

---

4 Run by the National Health Service.

builders, especially women builders: they wanted to learn more craft skills, so as to inform their own design skills: and they also wanted to overcome the idea that building sites were no-go areas for women, by understanding construction processes better.

Many addressed this by learning and practising a building skill. The Women Builders Co-operative was set up by some architects from the FDC with builders from Lambeth Women's Workshop. Links were made with Women and Manual Trades, which had begun as a support group for women builders and campaigned for equal access for women into the building trades, and later for special trade training for women. At least seven Matrix members did building work professionally and privately[5] and later produced a booklet encouraging young women to choose careers in building and architecture.[6]

The Feminist Design Collective split into two groups in 1980. Those who wished to take theory into practice set up Matrix Feminist Design Co-operative: others preferred to work within conventional practice structures and reform them from within.

Making space

American architects and historians Susanna Torre, Doris Cole, Gwendolyn Wright, Clare Cooper-Marcus and Dolores Hayden had written the only English language texts available in the early 1980s on women, architecture and the built environment, and provided the intellectual context for British feminists' ideological development. Their work in tracing the hidden histories of women in architecture in the USA was a corrective to the traditional formation most architects were undergoing in British architecture schools. Hayden's histories of utopian social structures linked feminism in history with other reformist and revolutionary initiatives in the production of the built environment, and amplified and deepened the feminist project.

In March 1979, the 'Women and Space' conference was

---

5   Fran Bradshaw, Sue Francis, Julia Dwyer, Janie Grote, Anne Thorne, Suzy Nelson, Elizabeth Adams.

6   Matrix, *A Job Designing Buildings*, 1980.

held at the Architectural Association, at which architects and anthropologists discussed:

[a] feminist critique of buildings and space which recognizes the importance of the social and political context of both the organization of production and the design process itself.[7]

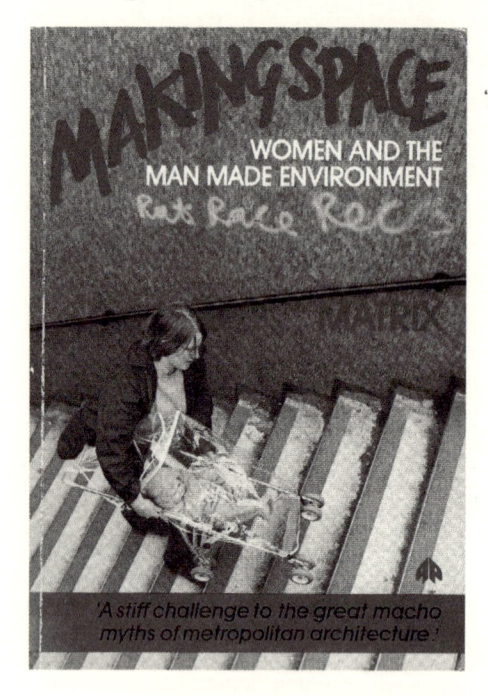

'Making Space', 1984

Some of the organisers of 'Women and Space' continued meeting, and in autumn 1980 formalised their discussions as a 'book project', leading to the publication of *Making Space: Women and the Man-Made Environment* under Matrix's name in mid-1984.

Influenced by contemporary UK and US feminist theories and practices which were evolving outside the architectural world, *Making Space* was the first British book to link these with the production of the built environment. The writers brought the notions that housework is also work, that women's history has been suppressed, and that it is necessary for women to organise separately from men, into their observations. Some important conclusions were that, as buildings

7   Sue Francis, *Heresies 11*, 1981, p.17.

and cities have been created by a dominant gender, they are man-made space and as such, are not neutral but expressive of social values and relations, that the domestic environment is a fit site for academic analysis on its own and in relation to public space, and that women's empowerment could come through collaboration and consultation.

Matrix feminist design co-operative

Matrix's design group began almost simultaneously with the book group and included three contributors to *Making Space*. The condensing of ideas which were fresh and still evolving into a book, and the experience of researching, writing and editing it, was able to give the new practice maturity and confidence. The research and criticism carried out by the book group writers instilled a practice of experimentation and discussion. The collective working methods set up by the book group were also to influence the new architectural practice, and vice versa.

Matrix was set up in 1980 in Anne Thorne's house in East London. From the outset, conventional professional practices were dissected, analysed and re-ordered. The analysis was not carried out in a laboratory or even within a school of architecture, but in a small co-operative business subjected to the usual demands of day-to-day practice. Working methods were developed which addressed women's inclusion in and exclusion from architectural processes and the built fabric of the city. Passion and commitment to women's causes, a belief that all the issues could be dealt with through altered practice, and a dose of middle-class guilt motivated the designers to identify the issues and begin to deal with them, one by one.

At the outset, some decisions were taken which were essential to forming Matrix's identity. The first of these was that it was a women's practice: the founders were not prepared to accept the compromises they had experienced working in conventional practices. More fundamentally, as the Matrix book group said in their Introduction to *Making Space*:

> We believe that, precisely because women are brought up differently in our society we have different experiences

and needs in relation to the built environment which are rarely expressed.[8]

This decision gave rise to discussions about the meanings of equality, and the contradiction of calling for equality of access to the profession while setting up a women-only practice, echoing arguments throughout the women's movement. A consideration of equality meant addressing the structural inequalities among co-op members. The co-operative would not have taken the shape it did had not some childcare been arranged and paid for. Awareness training on disability, racism and sexual orientation was arranged. Contracts of employment and equal opportunities statements were carefully assembled.

Its involvement in earlier critiques of the architectural profession led Matrix to adopt a 'no star' system, in recognition that architecture is a collective process. There were to be no individual names, and a corporate identity: the organisation was to be greater than the individuals who made it up, and it was expected to outlast them. The organisational model was neither a partnership nor a company, but a worker co-operative limited by guarantee. This ready-made structure was favoured by radical businesses, and ensured that profits stayed within the co-operative. Influenced by other feminist groups, Matrix chose to be eased into existence with the help of a Support Group of other architects and administrators over a transitional period.

There was little division of labour: it was only after time that administration and finance work became separate jobs. The relative status of architects and administrative workers within society reflected in their pay and conditions was dealt with by having an equal pay structure in which all workers, architects or not, could become members. Clients were to be found among voluntary groups, local boroughs, housing associations and co-operatives, not private individuals or companies. Architects could then work with people who do not normally have access to architects, specifically in Matrix's case, working-class women.

[8] Matrix, *Making Space*, p.7.

'Women architects: their work'

The historian Lynne Walker curated the historical section of the exhibition 'Women architects: their work', for the Royal Institute of British Architects in 1984, and edited the exhibition catalogue, which included her history of British women architects, published for the first time.[9] The exhibition was divided into two sections, historical and contemporary: Matrix was not included in the exhibition or the catalogue, as it did not present work for selection. The contemporary section included work by Mitra (some of whom had been a part of the FDC), women architects from Solon Design Group and Zaha Hadid.

Throughout its life, Matrix remained outside the RIBA, concerned that engagement with the RIBA would compromise the questioning that was evolving, and that the RIBA Women's Group was insufficiently critical of the context in which it was designing and building. The 90-93% male majority in the British architectural profession and the resulting inequality of access, pay and conditions were opposed by all feminists, including those within the RIBA. However, feminist architects were simultaneously developing a radical critique of the profession, and were uneasy with the reductiveness of the principle of 'equality': it was not sufficient simply to have more women enter the profession as it stood. Despite these qualms, the issue of equality of access was supported and addressed whenever it was raised, often by architecture students at the schools where Matrix had been invited to discuss its practice.

Projects

Parents and childcare workers, all women, had set up a local state-funded children's centre in Dalston, East London, and had acquired a former 'warm baths' building. Matrix had worked on Dalston Children's Centre's temporary premises and became their architects for the baths building conversion. Matrix soon discovered the shortcomings of using conventional architectural representations: their early drawing style was criticised for its aura of completeness which discouraged

---

9   Lynne Walker (ed.), *Women Architects: Their Work*, London: Sorella Press, 1984.

DCC from understanding the designs and developing them. Participative processes developed under the pressure of working with a mixed race and class women's organisation which gave high priority to access for people with disabilities. Drawings were rethought, and Matrix used a large-scale model, similar in appearance to a doll's house, to represent the building and to invite discussion.

The practice moved to a new footing when it was commissioned by an Asian women's organisation to design Jagonari Educational Resource Centre, an ambitious and high-profile educational centre for women from the largely Bangladeshi community in Whitechapel. Early discussions with the clients about the building design drew out issues of cultural difference as well as the impact of racism on the lives of the client group. At the outset, the clients wanted a building which would not appear 'Asian' on its very public front elevation as they did not want to invite abuse from racists or from men affronted by the idea of a women-only institution. Later, prompted by the 'Englishness' of the designs being presented to them by Matrix, they decided that the building's exterior should express some quality of Asianness. The group worked closely with the architects as the building was designed in

Jagonari Educational Resource Centre

detail, visiting exemplary buildings, tracking down bricks of the right colour and texture, mutually deciding on childcare requirements, safety, disabled access, social and educational needs, compromises between privacy, security and a sense of welcome in the entrance spaces, and so on.

Expressions of this collaboration are evident in various instances. The volumes divide into a 'parent' building and a crèche building separated by a courtyard, the créche protected by the larger building from the noise and pollution of the main road. The huge kitchen is positioned in a public, visible space next to the main entrance, and includes an area for traditional ways of preparing Bangladeshi food. The semi-colonnaded courtyard space and screened and deflected entrance sequence are informed by Islamic town house plans. The hybrid elevation design is influenced by Islamic architecture and the Arts and Crafts Movement: blue/green steel grilles over the windows refer to Islamic screening while providing security (they have been put to the test on a number of occasions since), while large areas of brickwork punctuated with a series of window openings and eaves descending over ribbon windows in the courtyard elevations are influenced by Webb and Voysey.

Many demands were made of the young practice's strategic and professional skills in running a large-scale project and dealing with recalcitrant, adversarial and occasionally supportive local authorities, consultants and contractors. New members who had appropriate skills joined Matrix and the shape of the practice began to change. Women working for community and housing groups began to hear of Matrix's work, and were attracted to it because of its consultative approach and its growing body of work. Projects were diverse and differently demanding, and were usually characterised by two things: their instigators saw them as unique and innovative; and they were publicly funded although individually initiated.

Jumoke Training Nursery, in the South London borough of Southwark, combined a full day nursery with a training centre for child-minders. A derelict factory in a council depot was converted into a 25-place nursery with a garden on the ground floor, and training space, offices and a crèche with play balcony and garden access on the upper floor, connected by a lift. The possibilities of employing women builders on Jumoke were explored, and it was discovered that most London women builders were individual tradeswomen: few, if any, women's building companies existed. The local adult training college agreed to support a training course for women builders, run by women trade trainers and Matrix. This did not lead to women builders doing anything other than some finishing trades on the building: however, the clients asked for and were given courses on the Building Process, which served as introductions to the experiences the clients would go through and the responsibilities they would face once construction started.

The building is enriched by ideas arising from the collaboration with the client, which included placing the kitchen at the heart of the nursery, visually connected with the play spaces; coloured glass in high windows projecting dappled colour onto the play floors, and mezzanine play areas scaled to child height to vary the children's choices of play.

The Calthorpe Project, which had occupied and transformed a derelict site in Kings Cross into a community garden, needed a timber-framed building to replace its temporary shed. The clients were experienced in involving their users in decision-making, and held two consultative picnic days to bring the design of their new building into public discussion. Matrix displayed site models and images of timber-framed vernacular buildings from different cultures including Chinese and South Asian, participated in open discussions, and had detailed consultation meetings with the Calthorpe Project's user groups. A design for a large-span timber-framed single storey building with mezzanine and an upper office covered by a great cedar shingled roof was developed, but Calthorpe was unable to attract sufficient funding for this building in the end and scaled down its ambitions, opting later for a self-built building on a different site. The project for a timber-framed building and the potential of working with women builders were realised later at a city farm in Walworth, South London, where a timber farm building was built by women builders.

Evaluating Matrix: note from inside the collective

Early in its formation Matrix had decided that it would only work on state-funded, social projects, and social housing became an important part of its work. A housing association which housed single people coming out of institutional care asked Matrix to design eighteen new flats for a gap site in South London. This was to be the largest project Matrix had carried out, and allowed the practice to develop ways in which housing spaces could be appropriated and used, particularly by vulnerable people. The eighteen flats are grouped around three staircases, so that each resident only has to deal with five other immediate neighbours. Each apartment interior has two main rooms of approximately equal size, so that residents can decide themselves how to allocate uses to the rooms. Matrix argued for access from all of the flats to outdoor space in the form of a balcony or an enclosed part of the main garden, as outdoor spaces have the potential for cheap and rewarding transformation. As the new tenants come directly from residential institutions, it is important that the interior communal spaces, i.e. the stairways and landings, are generous and well lit so as to allow privacy when necessary, but also to encourage spatial appropriation and chance encounter. Matrix designed elliptical top-lit staircases and large landings, which have been colonised by pot plants and bicycles.

During the 1980s, women's projects were supported by borough councils, and women's centre projects were proposed for many London sites; Matrix was involved in projects for Hackney, Brixton and Bermondsey. Of these, only one, Hackney Women's Centre, was built. The borough council had made a rundown shop building on Dalston Lane available, and much of the limited funds for building work went into repairing it before it could be converted for use: here the kitchen, built by women joiners, was at the heart of the social space, and as much of the building as possible was made accessible for disabled people.

The repairing theme was taken up again in Matrix's first arts project for the Half Moon Young People's Theatre, a theatre in education group which had been offered an extremely dilapidated nineteenth-century former town hall building in Stepney by the borough council as premises for a rehearsal and office space. This pompous municipal building posed interesting issues to the clients and the architects: should they respectfully repair the building at great cost, returning it to its original ornate Victorian state; or should they retain fragments and otherwise bring it into an entirely different use, one which was appealing to young people? The theatre group needed a rehearsal space that was naturally lit, highly secure and able to be acoustically sealed in the evenings in a building which was fully accessible for disabled people. Matrix completed the first part of a phased repair but was unable to carry out the second phase after losing a fee bid.

Technical advice for women's groups

Technical aid

In the late 1970s and early 1980s, voluntary groups and women's groups were setting up all over Britain; some of these groups approached Matrix for technical advice. Individuals within Matrix were already part of a discussion among socialist architects about the notion of state-funded 'technical aid', where design and technical advice was made available to voluntary organisations, just as state-funded law centres could advise groups and individuals about their rights. Matrix was an active member of the association of community technical aid centres (ACTAC), which represented technical aid in Britain and was funded by the government during most of the 1980s.

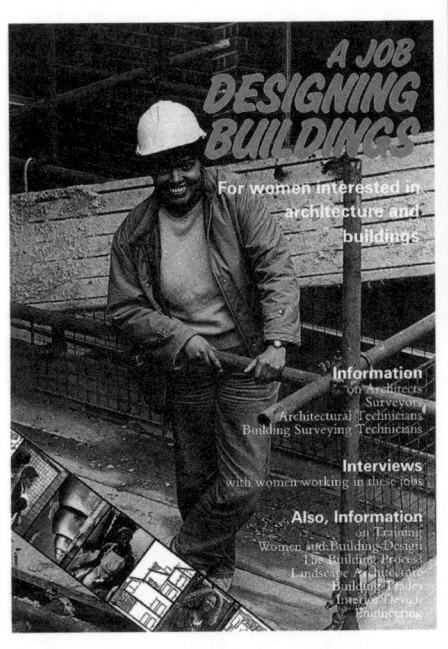

Above: Cover, 'Building for Childcare'
Left: Cover, 'A Job Designing Buildings'

A significant proportion of Matrix's work was technical aid, funded by the Greater London Council. Technical and design support was given to over eighty women's and other voluntary groups over seven years, and connected Matrix to groups throughout Greater London, from Earl's Court to Dagenham. The advice ranged from answering single enquiries to doing large feasibility studies. Two publications were supported. One, *Building for Childcare*, was written for community groups and childcare workers and it gave advice on the design of buildings for the under-5s. The second, *A Job Designing Buildings*, was intended for young women considering a career in architecture or building.

Women's Design Service, which shared premises with Matrix during the late 1980s, had also been funded to provide technical aid, but had changed direction in the mid-1980s and had become a research-based organisation. It was responsible for a range of publications including *Older Women and Housing Design* which was produced with some input from Matrix workers in 1991.

Education

Collaboration and participative practices involve the sharing of knowledge, skills and information: the Matrix designers recognised the need to be creative about the structures for this to happen. Consultative processes often created educational material which could develop into educational courses or become incorporated into publications such as *Building for Childcare*, which includes a chapter on the building process which began as a course for the Jumoke workers. A course on technical drawing, which started as a consultative tool for Dalston Children's Centre, was developed further for use on women builders' training schemes, particularly at Women's Education in Building (WEB) in Notting Hill. Some of this work helped develop teaching material for an access course for women into architecture and building (WIAB) which was set up by Yvonne Dean at the University of North London and was repeated at South Bank University by Marina Adams. Several Matrix designers contributed to this course.

Matrix was often invited to speak at schools of architecture, sometimes by teachers and other times by students, in Britain and abroad. Other courses were more diverse: urban studies centres invited Matrix to devise events for schoolchildren, and women's and housing organisations sent invitations to contribute to their courses and conferences.

Networks

Every year dozens of women architects and architecture students visited the practice to discuss its work and their position within architecture. This connected Matrix to a wider network of radical women architects who met regularly and set up two events about specific issues: 'Women in Architectural Education' for teachers and students, and 'Feminism versus Professionalism'. The 'Women's Realm' conference, which attracted about 200 architects, students, teachers, artists, builders and gardeners to a cross-disciplinary event at the University of North London in 1988, was the last of these events, but the network revived under a new name and met again during the 1990s and into the new century, helping in the birth of twenty-first-century initiatives such as Taking Place.

Poster for 'Women's Realm', 1988

Looking back: discussion, 1999

Some of us met again to make some interim conclusions about Matrix and to relate it to contemporary practices. We ended our discussion with more questions. Something we discussed deeply can be loosely described as the prioritising of process over product which occurred in Matrix's practice. A building design is mediated by gender and social relations right down to the detailing and the materials chosen; although we were aware of the social and political influence expressed in relationships in plan and space between activities within buildings, which came initially from feminist studies of the design of domestic spaces, we did not often subject our design work to the same level of scrutiny that we did other processes and so were unable to develop this further. We had not yet had the years to develop a language to discuss design issues that was independent of the normative language we had imbibed during our education. We did have a language for the analysis of the building process and consultation, partially taken from the Women's Movement. What is interesting is that a politicised language of design analysis has now emerged.

Although Matrix was not a campaigning group, but a practice devoted to building, its influences were disseminated by publications, videos, lectures, teaching, courses and contacts with women students and practitioners. Was this in the end more important than the buildings themselves?

Many creative moves were made in a very short time: energy was gained from confidence in and commitment to political processes. The political climate when Matrix appeared was one of radical re-ordering in which many things were being tried out. A continuous thread was the question: who are the people we are designing for? The answer was often expressed in terms of a type or group, e.g. 'women', 'disabled people'. Today, there is a focus on the individual experience of oppression. This legitimises designing only for oneself, and no one else. There is a corresponding lack of confidence in drawing conclusions about conditions and oppressions. While this is an understandable outcome of the fragmentation of politics of which we were a part, is there a failure in current practices to deal with inequalities?

Contemporary practice

Matrix ended in the mid-1990s. Our current work as architects and teachers has been informed by the intense analysis of the Matrix period. The work we are now doing connects back to the early explorations through its interest in the processes of designing and making buildings, as well as in the end product, in working with clients involved in social projects, in our continued connections with other women architects, supporting their work and promoting evaluations of theory and practice, and in our openness to working across disciplines, with artists, writers and other makers.

Chapter 4

muf

AN INVISIBLE
PRIVILEGE

Ingrid Ruudi

Chapter 4   muf

SLE

Danny Dorling

AN INVISIBLE PRIVILEGE 59

muf's[1] text for the *Alterities* conference is now something of a historical document which can be found strewn through *This Is What We Do: A muf Manual* (2001).[2] For this book, we enclose an account of a piece of work made in London in 2004 and an interview with Ingrid Ruuidi, an Estonian architect. The afterwards is by Juliet who originally spoke at the *Alterities* conference: *you don't have to be in the same room to continue a conversation.*

An invisible privilege...

In summer 2004, muf architecture/art were invited to contribute a piece of work to the Architecture Biennale in Clerkenwell. *You probably were not invited to come along.*

Ten years ago, muf was loaned the top floor of a building on Great Sutton Street for six months, 2000 square feet of rentless space that enabled muf to move from kitchen table to studio and set up as a viable, working practice. Ten years later, muf finds itself in the same place again *(53-56 Great Sutton Street)*, invited to make a work for use of the ground floor now... borrowing space to make work in again.

muf was founded in 1994 with a commitment to work in the public realm critiquing the confinement of care and feeling to the domestic realm (as though lives and feelings could be left behind at home). Though in reality they never are, what gets talked about in the studio is always already a personal slanting of 'professional' attitude: the cost of running the studio, of making good work and money, the costs of working in the public realm, of childcare, of staying away from home, to have children or not... A personal slanting of professional judgement that continuously tests a larger, ongoing discussion of the limits and possibilities of creative practice: how do you find time and make space (literally) to make good work?

---

[1] muf is a collaborative practice embracing art and architecture. The practice was formed in 1995 by Juliet Bidgood and Liza Fior and artist Katherine Clarke, with the express intention of working in the public realm. Though Juliet Bidgood left muf in 2002 to pursue a parallel career, she remains an occasional collaborator. The urban theorist, Katherine Shonfield, though never formally a partner, was also a key collaborator from the outset until her death in 2003.

muf actively supports flexible working. This has proved attractive not only to those with children but also to those involved in teaching architecture and in part-time study of related disciplines. In the past ten years 44 women have worked at muf and currently there is one full-time male employee.

[2] Published by Ellipsis in London in 2001.

> muf's practice has always been informed by this doubling of professional-personal knowledge, where the personal is always already public-political. There is a basic gender-consciousness in this, an acute appreciation, as an all-women practice, of how the professional is always already private, how public space and private time are contested, how public, creative ambition seems to get split arbitrarily across gendered lines. Room for creative manoeuvre becomes a privilege, dependent on a fortunate coincidence of timing and geography. Where you are, where you find yourself, determines the possibility of what you can do and when: *53-56 Great Sutton Street, a room of your own.*

Geography matters to the extent that it is as important as history in providing the context through which people's lives are led and influences the courses those lives take. It is not a variable just as time is not a variable.[3]

> *Before muf there was Matrix.*[4] Historically, there has been a tendency to tolerate only one female practice at a time, as though there were limited room and life span for only a single grouping of women.
>
> Women tend to be compared to other women, leaving no room for level-term comparison within the mainstream. For muf, this created the dilemma of taking up the discourse as an all-women practice, as a singular space in the mainstream media in the UK, and how this projection of us as a rarity, in turn, tries to contain the content of what we were and still are actually saying or doing… This is why perhaps we have chosen more and more to make our work speak for us.
>
> It seems ironic from this position that a women-run practice is invited to exhibit itself in a male-dominated Architecture Biennale. We could easily have made use of this as an opportunity within the promotional terms of the Biennale: an opportunity for self-exhibition, for containing

---

3   Danny Dorling, 'Neighbourhood Impact – seeing is believing', paper for *Analysing Neighbourhoods and their Impact* Conference at Clifton Hill House, University of Bristol, on 15-16 June 2004.

4   *Alterities* was the first time muf had spoken at a conference centred around feminist theory and also the first time they had shared a platform with Matrix.

An invisible privilege

and completing 'finalised' work. But after ten years of practice, finding ourselves – by coincidence – back where we started out, this was a chance to reflect on the conditions that made muf's creative practice literally possible, those seemingly arbitrary privileges and coincidences of time and space: *2000 sq. feet of rentless space loaned to us back in 1994...*

53–56 Great Sutton Street. Ground floor. The full face of the back wall 15m by 3m, is painted over with a series of statistical markers and maps. The newly-whitened space spoiled with a layered wall painting of coloured, descriptive maps that reveals with added layers over the course of the week a gradual accretion of social knowledge, of the gendered lines of privilege.

Saturday 19 June
A base map of London shows the proportion of households employing domestic servants in 1911. The historic blocks of wealth and privilege are lined out, made visible. The map spills onto the floor.

Monday 21 June
87 blue spots are added to the base map. Each spot marks the existing location of a blue plaque awarded to a 'notable' woman... Mary Kingsley, Florence Nightingale, Anna Freud, etc. – the site-specific markers of historic success.

Wednesday 23 June
A new layer of paint is applied: a pattern of hexagons marks out the incidence of degrees awarded across London (describing a contemporary topography of educational achievement).

The layers mark out a genealogy of privilege, a historic correspondence of privilege and cultural 'successes' that are geographically aligned. The highest incidence of awarded degrees overlays and corresponds to the area with the highest proportion of domestic servants from a century before. The blue-plaque markers of 'successful' women, in turn, overlay these sites of privilege almost seamlessly. The possibility of creativity (of making room for creativity), the possibility of success seems dependent on some sort of hidden spatial matrix of social circumstance... as though there were invisible lines of cultural inheritance here, correlations of privilege that are only made visible here, when they are laid out explicitly as a map, on a wall.

muf's work moves across these lines of privilege and reward, shifting between high-profile establishment settings from the Tate Gallery or a Biennale to other less 'visible' settings, to those projects and places that are left unspoken about: Tilbury, the Greenway, Rudolf Road... places that sit outside the discourse of celebratory design, projects that would not fit within the blue-plaqued sites of privilege and success.

280 invitations (screen-printed on heavy card) are sent out to all the women who work in the architecture practices within the Biennale area (architects and administrators alike). An invitation ensures a glass of champagne and the opportunity to read the growing layers of this map on the wall at a private view that lasts a week.

An invisible privilege

Representation in the Architecture Biennale was always going to be limited. The Biennale itself is a showcase of that architectural talent, a moment of highly localised professional self-exhibition (of a profession that is overwhelmingly 87 per cent male). muf's women-only invitation to an invisible privilege was meant as a provocation, testing the tolerance for women, for all that is Other within this 'establishment' setting: how much room can you create for what is more easily suppressed, left unmentioned?[5]

53-56 Great Sutton Street, an invisible privilege

An interview [6]

INTERVIEWER: *How did muf start out as a female practice? Was it a deliberate choice and aim?*

MUF: It began with the ties of friendship – friendships which began in three institutions. Juliet and Liza met at the Polytechnic of Central London recognising each other as fellow travellers and students in a structures lecture. We

---

5 The majority of architecture practices in the Clerkenwell Biennale area only offer the statutory *minimum* of maternity benefits. The statutory *minimum* of maternity benefits from employer is six weeks, which is a *minimum of tolerance.*

6 This interview with Ingrid Ruuidi was previously published in the Estonian journal, *Ehituskunst.*

went on to work together, burnt out and stopped until muf. Katherine Clarke and Liza met at the Architectural Association where we were both working and went from laughing in the library to teach together. Katherine Shonfield and Liza met at the RIBA having known of each other only through distorted male reports. Each friendship was fuelled by the way that each institution did and did not accommodate our presence.

We started working together at first as a loose grouping and then as muf, because of our shared interests in working outside our disciplines and the expectations of practice. I was pregnant at the time with my first child and utterly aware that if I did not create a context to work in, I wouldn't work again.

You have been referred to and discussed from gender-centred points of view. Perhaps not so much now, but when you started out, gender-consciousness was a heated subject in theoretical discourse. How do you position yourselves in that discourse?

Before muf there was Matrix. The architectural press reminds us of our lineage, as if there can only be one female practice at any one time – like an insect that dies as it gives birth. Women tend to be compared to other women. As soon as you forget your gender, you will be sharply reminded of it by men. As we move into our forties, gender-consciousness is a consciousness also about choices – children and not children, making good work and making money, the cost of a studio, of housing in central London – all gender-related and debated within the studio.

Do you feel theory to be overestimated in architecture?

No. The build and building the built take such a long time and often do not result in an object, rhetoric can fill that gap. It can make a bridge between – or leapfrog over – one worked-out project and the next. The retrospective description explains a piece of work not only to others but to yourself. Since we stopped teaching, theory has been intimately linked to practice.

You have always stressed practice as a form of research and as a basis of design. Your method seems to be positively instrumental. Does that mean always starting from scratch?

We build on what we know – the world is infinitely more complex than your understanding of it. We begin from the premise that any solution is flawed. It is just about getting it as right as you can. The method is a combination of the dogged surfacing of the hidden and the obvious structures and interests that make up a given situation combined with wilful intuition.

I would describe your idealism and flexible ego as feminine qualities... but how far can you go with collaboration? How do you achieve balance between client's wishes and not losing yourself?

How nice that you don't instead perceive loads of ego but a flexible idealism. There is no loss of authorship in building a relationship. It does seem to be a prerequisite that for a project to be successful we have to like, empathise with or take an interest in the client body. We haven't really had an example of creative exchange as an act of revenge. In a recent studio crit, a well-known male urbanist was unable to comprehend the value we ascribe to negotiation, as if that degrades the object out-come – the completed built project – rather than enhances it.

You seem to deal a lot with the ordinary, the minor, and the local – all pretty much overlooked in Grand Architecture?

Yes, in part, because the types of places we often work in are themselves ordinary, local or minor. But, in fact, close looking is always combined with a moment of import. When we designed a park in Tilbury, east of London, we started by noticing that there were horses kept adjacent to this housing estate, so we organised a gymkhana. It was a highly formalised way for horses to occupy the site, a conceit that legitimised their presence for an afternoon. The move was then concretised by the design for the park itself.[7] We have worked in 'grander' locales, e.g. designing exhibitions for Tate Britain where we apply similar processes starting with the paintings themselves.

Consulting children seems especially unusual. Do they make a good partner and a good mirror of social reality?

Yes. New pedagogies value the informal education that is gained through experience. It's a good model for practice, that idea of a wakefulness where everything matters... just as you

---

7   cf. muf, 'Rights of common: ownership, participation, risk', in P. Blundell Jones, D. Petrescu and J. Till (eds), *Architecture and Participation*, London: Spon Press, 2005, pp.211-16.

watch the literal process by which children become social beings, which is, as you say, a sort of social mirror.

As women are traditionally associated with the private sphere, you deal almost entirely with public space. Does contemporary public space lack sensitivity?

The crude expanse of an airport runway has its own pleasures. But, yes, we do strive to make spaces that combine largesse with intimacy, a simplicity, a dreaminess, rather than the cloying, anodyne formulas for 'lively' civic spaces.

Oysters wall, St Albans Hypocaust Pavilion, 2005

Some of your projects are very tactile, somehow close to the body – the 'Purity and Tolerance' installation comes to mind. What does the body and space relationship mean to you?

The world is continually experienced at multiple scales – the vista and the proportion of cotton to acrylic in your vest, from the micro to the macro, from detail to strategy and back again. Immediacy, proximity, detail, the tactile.

Whereas the projects themselves are very well grounded, based on clear conditions, you often add some extra message or a key with the title: pleasure, desire, dream, seducer recur in the rhetoric. How do you conceptualise 'desire' and 'pleasure' in relation to architecture?

'Pleasure', 'desire', 'dream', 'seducer', 'the borrowed', 'the lent', 'the cared for' – all these operate alongside all the things that are supposed to govern us. Alongside but not mentioned.

Afterwards

The history of modern British town planning and urban design is as much concerned with the un-built as it is with built form. For example, the particular quality of London is determined by its greenbelt and royal parks and, in these spaces, by the contested territory of recreation and leisure. It was the collective assertion by ramblers to rights of recreational access to land usually reserved for hunting that informed the policy of not building, which created the greenbelt and defined the physical framework of the city.

muf's continued assertion is that there are preconditions that need to be met in order to make buildings worthwhile and that we should take care not to unwittingly build in limitations. This text (written five years ago) sets out the method by which those preconditions can be explored and established. Importantly, the processes of creating public spaces outlined here require willing and able partners, clients who are themselves open to engaging with their public/their audiences and open to collaborating in a shift of expectations.

The afterwards – reference to the personal (highlights a frustration but also) – points to the understanding of the personal as political and the expansion of this principle as a method of brief development. As the footpaths marked out by the ramblers' desire(s) to leave the city at the weekend, which in turns skirted, criss-crossed and staked out what were to become shared public spaces, so, too, the detail of the woman accessing the museum with her child, the child walking to school, the necessities of using water and the making of ceramic – along with visual and other pleasures – inform the brief, specification and fabrication of the built.

Chapter 5

Jane Rendell

Luce Irigaray

HOW TO TAKE PLACE
(but only for so long)

Jane Rendell, Barbara Penner and Iain Borden

Matrix

Teresa Hoskyns

Chapter 5   Jane Rendell

Beatriz Colomina

KE

Hélène Cixous

H.D.

HOW TO TAKE PLACE (but only for so long) 71

Take 1

In June 1999, I gave a paper at a conference called 'Alterities: Interdisciplinarity and "Feminine" Practices of Space' in Paris, organised by Doina Petrescu. The delegates included men, but mainly women, from France, the UK and the USA, both older and younger generation feminists, as well as theorists and practitioners. Interestingly it was the USA and UK feminists who referenced the French feminism of the 1970s in their work, while the French contingent, particularly the older architects, seemed to oppose these writings. They felt that, as women, they had been excluded from the architectural mainstream in France and that theory had not helped. While I agreed with their position in some ways, I was not prepared to abandon theory, at least, not yet.

During the short time I had worked for the feminist architectural co-operative Matrix in the early 1990s, we talked a lot about feminism, but there were never any conversations about aesthetics, and certainly no theory. It seemed to be beside the point. A decade later, muf, the all female art-architectural practice who seem to have succeeded Matrix, now stand for all the 'f' words in architecture. Yet muf do not describe themselves as feminists. Even their name signifies a fundamental shift in feminist politics, from an earnest modernist heart-on-sleeve approach to a post-modern sexy playfulness. muf consider the way that they work, the very processes they adopt, to constitute the 'form' of the work, to provide the aesthetic content.[1] And it is this concern with an aesthetics of process that connects muf's work to French feminist theory and so to feminism.

At the Paris conference, members of muf and Matrix were placed on the speakers' platform at the same time. For me this was a vital moment: where I could see that a new approach to feminist architecture was required, that engaged ethics and aesthetics, what Doina has called 'poetics and politics'. The majority of women in 'Taking Place'[2] (see Take 3) were

---

1 See, for example, Matrix, *Making Space: Women and the Man-Made Environment*, London: Pluto Press, 1984, and *This Is What We Do: A muf Manual*, London, Ellipsis, 2001.

2 'Taking Place' was a name the group chose for itself after a text by Teresa Hoskyns.

participants at that conference – a significant event that, like *Sexuality and Space* and *Desiring Practices*, I believe, has marked out a new stage in feminist architectural practice and theory, one which describes itself, not as feminist, but in terms of the 'feminine'.[3]

During the conference, something else changed for me, on a much more personal level. The conference was arranged as a series of papers delivered over two days. In the mornings of both days, papers were given in 'amphitheatre 1', and in the afternoons there were two parallel sessions, one in 'amphitheatre 1', the other in 'amphitheatre 2'. My paper was placed in an afternoon session, in 'amphitheatre 2' with five other speakers, three of whom were my students, each of whom I had been tutoring in preparation for the conference. A male colleague from my institution chaired our session.

I felt for the first time a very real conflict between my theoretical knowledge underpinned by my political aspirations, and my emotional state. Up until that point, although it was not yet the subject of my writing, I believed strongly, from a feminist position, that teaching and writing should be conducted in ways that challenged the vertical power structures favoured by certain forms of patriarchy. Women, located on the outside of these structures, were in an excellent position to create alternative horizontal networks, where relations and dynamics of power could be constructed differently. Further, in opposition to the laws of property owner ship that capitalism holds dear, I was an advocate of a feminist economy of gift, an economy which valued interaction, reciprocity, the free exchange of ideas and the emancipatory power of giving.

It suited my theoretical position to deliver a paper alongside my students. Yet, it felt difficult to do so, because this meant I ran against the grain of the conference, which despite the content of the papers, in form bore similarities to certain hierarchical aspects of power that I associate with patriarchal capitalism. The position of the male chair remained intact, for

---

[3] Beatriz Colomina (ed.), *Sexuality and Space*, New York: Princeton Architectural Press, 1992, and Duncan McCorquodale, Katerina Rüedi and Sarah Wigglesworth (eds.), *Desiring Practices: Architecture, Gender and the Interdisciplinary*, London: Black Dog Publishing, 1996.

example, as did the two-tier conference structure, the main morning sessions with the well-established speakers and the split afternoon sessions consisting of speakers whose work was less known. The relation of tutor to student also seemed to follow a pattern. In the morning sessions, tutors discussed the work of their students in their own papers, whereas in the afternoon sessions, tutors (or perhaps I was the only one) presented work alongside their students.

I had been planning to deliver a version of 'Undoing Architecture', a piece I had written concerning feminine rhetorics, my term, to describe an undoing of architecture.[4] The talk I had prepared combined three voices: the male architect, the feminist theorist and the voice of the unruly user. When my paper, along with those of three of my students, was accepted, in preparation for the conference, my male colleague set up a situation where I found myself 'rehearsing' my piece alongside them, while also offering advice on ways to improve their presentations. Interestingly, and perhaps as a result of my response to the structure of the conference, when I arrived at the venue, and experienced the place and my position within it, I decided to change my paper. At the last minute, I had the introduction I had written to a book entitled *Gender, Space, Architecture,* which I had co-edited with Iain Borden and Barbara Penner, faxed from London.[5]

As I listened to the sound of my voice, I realised my last-minute change had been a rash one. My editorial spoke with the voice of authority; it claimed to lay out the territory of feminist architectural theory and practice over the past thirty years. It was a voice that might be described as masculine, it was clear and knowing, which 'placed' the work of many of the women present, including those who had been speaking in the other auditorium, according to my 'world-view'; this was different from my initial paper, with its series of complex interactions, composed of multiple voices, spoken very much in a feminine mode.

4   Jane Rendell, 'Doing it, (Un)Doing it, (Over)Doing it Yourself: Rhetorics of Architectural Abuse', in Jonathan Hill (ed.), *Occupying Architecture*, London: Routledge, 1998, pp. 229-46.

5   Jane Rendell, Barbara Penner and Iain Borden (eds.), *Gender, Space, Architecture: An Interdisciplinary Introduction*, London, Routledge, 1999.

My reaction to what I perceived as an act of marginalisation was to set out my own framework as an oppositional move. Upset, and taken over by emotion, I had ended up playing the same game, using the language of the master I had originally sought to overturn. I had taken up the 'master's tools', as Audré Lourde would have it. This was a language that placed others, many of the women present, within my own world vision, where in order to restore my own equilibrium, I accorded myself the very position of power and control from which I imagined myself excluded.

To share thoughts, as I do, when they are still burgeoning and when the imprint of publication has not yet placed them as one person's property, produces much confusion over the ownership of ideas and questions of acknowledgement. If I reject the values of patriarchal capitalism that I believe run right through academic research and the institutional power structures that support this culture, then it is really no good when feeling marginalised to seek solace inside those same constructions, by creeping back inside or by making a similar space from which to speak. But, on the other hand, I can no longer accept, at least uncritically, a model of the feminine economy as one of self-less exchange. There are some serious flaws, I now realise, in the theoretical assertions of the gift economy in French feminist theory, particularly, but differently, in the work of Hélène Cixous and Luce Irigaray.[6] Does the gift economy really critique the asymmetrical power structures of patriarchy? Does it not construct other asymmetries, ones where the giver, by putting herself forward as self-less, accords herself, through a perceived altruism, a position of moral superiority? What about the expectation the giver has of receiving a gift in return? Or the pressure placed on the receiver to reciprocate? And worse still, what is it that happens to both when no such gift is forthcoming? (It is fair to add that Cixous has acknowledged problems with the gift: that what the gift offers is not simply giving, but the possibility of establishing new forms of exchange, ones that do not follow models

---

6   See, for example, Hélène Cixous, 'Sorties', in Elaine Marks and Irene de Courtivron (eds.), *New French Feminisms: An Anthology*, London: Harvester Wheatsheaf, 1981, pp. 90-8, 96-7 and Luce Irigaray, *Elemental Passions*, London: The Athlone Press, 1992.

of self-same and accumulation.) And then, there is Derrida's suggestion, slightly pessimistic I used to feel, that the gift is impossible, that all the gift can give is 'time'.

It is important to imagine new models of reciprocity, new two-way exchanges, patterns and dynamics of equivalence. For me, it is vital to move forward and seek out ways of working and exchanging ideas that develop mutual respect and responsibility. An exploration of these relationships in terms of ethics and aesthetics is inherently concerned with processes that are spatial, temporal and material, as well as cultural, political and psychological. This is what my research is now concerned with: an investigation of the psychic and material space of such encounters, of what happens when one relates to another through 'giving'. I call these writings 'confessional constructions'.[7] It is here that theory can still help, in re-conceptualising the spatial dynamics of new relational forms in terms of positionality and subjectivity. But it is also here where theory alone produces a state of illusion and expectation. In my experience, it is only by acting out these conceptual forms, by practising them, that processes of transformation and realisation start to occur, often difficult, sometimes too difficult to bear.

Take 2

Just over a year after the Paris conference, in August 2000, I was invited to participate in the International Women's University, an extraordinary event that took place in several German cities over the summer months. I was located for a week in the town of Kassel, giving lectures and seminars on work I had published on the topic of gender, space and architecture. I talked about the relationship between feminist theory and architectural space, in particular about the spatial ways in which we write and the complex interaction between theory and practice, in what I call 'feminist spatial practices'.

During the one-hour lecture sessions, held in raked traditional auditoriums with the lights down, it was not clear to me how well I was engaging with the audience. The diversity of

---

[7] See, for example, Jane Rendell, 'Writing In Place of Speaking', in Sharon Kivland (ed.), *Transmission: Speaking & Listening,* Sheffield: University of Sheffield, 2003, pp. 15-29.

the women who attended was incredible, both in terms of geography, where they came from, and also in terms of discipline and practice, what they did. There were women working for NGOs in India, architects from Brazil, sociologists from Nigeria... I was an English-trained architect and academic. Despite my history, having lived in many countries, in the United Arab Emirates, Afghanistan, Ethiopia and Sudan as a child, I felt uncomfortable.

Who was I to speak to these women? What could these intelligent people with such a range of experiences learn from me? And how was I to make a connection with them?

At times, the way I teach involves risk taking; it involves a degree of admitting 'not-knowing' and refusing to provide answers even if I have them to give. I love questions, to travel to the edge of what I know and, to paraphrase bell hooks, 'to meet my students there'. My love of teaching is, I think, generated through a genuine desire to connect with others in a reciprocal encounter.

With this in mind, I decided to carry out an experiment for my last lecture session. During the coffee breaks and lunches of my week in Kassel, I had got to know several of the women there fairly well, well enough to feel some sense of trust. So I took the paper I had been about to deliver and cut it up into pieces. The lecture I had been intending to give had been entitled 'Undoing Architecture', composed as in Paris, of three voices: the voice of architectural practice, the voice of French feminist critical theory, and the voice of a story teller – me – describing the taking apart, through unconventional DIY, of a house I had once lived in. Once again, just before the lecture, I decided not to deliver it; I felt uncomfortable with the authorial nature of the composition. So I asked for a pair of scissors and cut the paper up. I handed out the voices to all the women in the audience and asked them to take up any position in the raked auditorium they wished. Then I asked them one by one to read aloud the piece of writing they had been given, but in their own mother tongue.

What did I expect – a delirious cacophony, a rich celebration of cultural diversity, an overturning of the lecturer and lectured-at relationship? How wrong I was. I hadn't anticipated how fearful the students would be. Slowly people did start to read, but in English, in quiet and reverend tones, struggling to pronounce the words just right. While it was beautiful to listen to so many female voices, filling the lecture space rather than my own, despite the different tongues and intonations, the words were still mine. I could hear only my own writing ringing in my ears. I had failed to turn things around, to set Western academia on its head, but rather, in some strange way, I had inadvertently reinforced my own position of authority. How could these women speak, when I was silencing them with my own words?

What had been created in that room, was what I have come to call a 'confessional construction'. My attempt to create a participatory piece at least made me aware that stories of the self, confessionals, are not revelations but constructions. The confessional is a form of physic architecture.

Take 3

The conference in Paris brought together a number of women, theorists, artists and architects, including myself, who created an informal group. We called ourselves: 'Taking Place'. We started out by meeting in various cafés to talk, to eat and drink, but these were private discussions. Then we decided to go public and to turn ourselves inside out. For our first event at the University of North London in November 2001, I performed a 'confessional construction', a spoken version of the text 'Travelling the Distance/ Encountering the Other'.[8]

Prologue:
'confessional constructions'
Site:
at the foot of the stairs in the entrance lobby
Action:
to stand still and speak
Words:

[8] Jane Rendell, 'Travelling the Distance/Encountering the Other', in David Blamey (ed.), *Here, There and Elsewhere: Dialogues on Location and Mobility*, London: Open Editions, 2002, pp. 43-54. All words sections are taken from this piece of writing.

The confessional is not revelation but a construction of the self. It is a form of physic architecture. 'Confessional constructions' are a walk that takes us through the architecture school of University of North London, exploring the tensions and ambiguities that exist between the personal and the public, the autobiographical and the professional, academic thinking and emotional subjectivity in feminist spatial theory and practice. This walk consists of a series of critical spoken interventions at five specific sites in the building, encounters between subjects and sites that engage the intersection of personal reflection and institutional space. This walk will last one hour, starting now and finishing at noon.

> My practice, my writing, are concerned with the ways in which we construct relationships with others without losing ourselves. I call these writings 'confessional constructions'. This one was originally written for David Blamey for his book *Here, There, Elsewhere*. It is entitled 'Travelling the Distance/ Encountering the Other'.

> The distances we travel are physical and psychic, emotional and mental. The others we meet on route take the form of places, objects or people. They may be our teachers, critics, students, lovers, children, parents or friends. Most often the distant other we encounter in our travels is what we thought to be a familiar part of ourselves.

> *...we are voyagers, discoverers*
> *Of the not-known,*
> *The unrecorded;*
> *We have no map;*
> *Possibly we will reach haven,*
> *Heaven.* [9]

Intervention 1:
feminist academics come with baggage

---

9   H.D., *Tribute to the Angels: Trilogy*, London: Carcanet Press Ltd., 1997, p.59.

Site:
stairs from ground to first floor
Action:
to struggle up the stairs loading a bag with feminist texts picked up from each step
Words:
Gloria Anzaldúa
Frances Bartkowski
Rosi Braidotti
Judith Butler
Hélène Cixous
Diane Elam
Elizabeth Grosz
bell hooks
Luce Irigaray
Caren Kaplan
Julia Kristeva
Doreen Massey
Elspeth Probyn
Gillian Rose
Janet Wolff

Moving is not strange to me; both physically and emotionally, I am most comfortable in motion. For me, being in motion itself provides a sense of stability, having left but not-yet-being-there. And it seems I am not alone. Postmodern feminism is full of stories of travel.

For those concerned with issues of identity, spatial metaphors constitute powerful political devices and critical tools. Positionality provides us a way of understanding knowledge and essence as contingent and strategic – *where* 'I am' makes a difference to what I can know and who I can be.

Many of these women have not moved, but been moved. Much of their written work speaks of displacement. Movements vary in their political dimension. Not all journeying is to be celebrated.

First in one place then in another, I find it is easier to make connections with another when I am out of place. I like to take my baggage across the frontier into a new land – to unpack among strangers. Only to find all the things I have brought, have lost their intended purposes.

This also describes my experience of collaboration and interdisciplinarity. Familiar words, but ones which suggest different, yet equally complex dynamics about the relationship between two – working between disciplines, working with another. Many female academics I know work between disciplines. They seem to be drawn to the strangeness of new areas of study, to making relationships – connections and distinctions – between things. What travel and interdisciplinary study have in common is their potential for transformation.

To be interdisciplinary, you are between two places. The conceptual space of critical theory both sustains and inspires me. It allows me a place to reflect on the past and to imagine new futures.

Only by acknowledging the work of earlier feminists, can we operate 'behind', adopting ways of working that critique those who have gone before. Only by going forward can we imagine a world as a yet-unrealised female subject. Only in this state of mind, between past and future, can we open ourselves up to encounters with the other. We travel the distance to transform as well as transgress.

Intervention 2:
the games we never played
Site:
the pigeon holes on the first floor
Action:
to take cards out of pigeon hole marked 'R'
Words:
PO Box 1606, Riqa, Dubai, Trucial States, Arabian Gulf.
PO Box 387, El Fascher, Darfur Province, Western Sudan.
PO Box 1570, Kabul, Afghanistan.
PO Box 86, Mekele, Ethiopia.
22 Wells Close, Harpenden, Herts, AL5 3LQ.
15 Mayfield Road, Girton, Cambridge, CB3 OPH.
22 Collegiate Crescent, Sheffield.
15 Filey St., Sheffield.
3 St Quintins Avenue, London, W10.

12 Mogg St., Bristol.
22 Barony St., Edinburgh.
18 Eyre Crescent, Edinburgh, EH3.
60 Lillieshall Road, London, SW4.
2a French Place, London, E1 3JB.
48 Enfield Cloisters, Fanshaw St., London, N1 6LD.

    I was born in Al Mahktoum hospital, Dubai, in the 'Middle East'. As a girl, I lived in Sudan, Afghanistan and Ethiopia. My movements followed the pattern of my dad's work. Unlike many children in similar situations, I was not put into a boarding school at the age of 11, but came back to live in England with my mum and sister. I say, 'came back'. The phrase implies that I was coming back to somewhere I had already been. But I had never lived in England before. It was my parents' country of origin, but not mine. I never felt at home back (t)here. But I have never felt at home anywhere.

    Once the women were back at home, my dad continued to traverse the drier areas of the globe. He is a hydrogeologist. A man who looks for water and brings it to the surface for people to wash and drink. He does this in lands that are not his own, that he was not raised in, that are strange to him.

    It's only recently that I have come to acknowledge the difficult relationship I have with my dad. Secretly, I have always been rather relieved in confessional conversations around the academic dinner table that my relationship with colonisation is with acts of kindness. My dad was in a position to help, to use his knowledge to provide water. He is a gentle and unassuming man. So why am I uneasy? Because, generated though a sincere motivation to help, my dad's particular brand of colonisation was paternal.

    I remember nights spent capturing insects on cold stone floors. Our house was not grand, but unlike many other houses we had lived in, it had a stone floor, running water and a tin roof that didn't leak. While my parents were out, a Tigreanean man stood at our gate. He was our watchman. I was uncomfortable around this tall black man with his hardened feet and long white robe and stick. My younger sister played with him. But I felt distance from him. He never spoke to me. Nor I to him.

At that time all Westerners had guards positioned at their threshold. Was it because a Western family had recently been ambushed nearby by Tigreanean rebels? Or was it to suggest that we were important enough to protect? I am still embarrassed by the fact that Africans looked after us and our house. 'Why?' I ask my parents now. It was the custom, they say.

If it is a custom, it's one that shames me. I never played with the watchman. I never travelled that distance. Do the games we never played make me a coloniser too?'

Intervention 3:
prepositions – transformational messengers
Site:
first year crit space on first floor
Action:
to lie, sit and stand at different positions in the room
Words:
At
Through
In
Between
With
Among
On
Under

A while ago I went with a group of women to hear French psychoanalyst and philosopher, Luce Irigaray, talk at the Architectural Association in London. Irigaray spoke of her early research into language with 8-year-old Italian girls and boys; that when given a preposition to use, girls made sentences that linked them to people, whereas boys made sentences that linked them to objects.

There were three questions from the floor to Irigaray. All from men. All angry. The first asked her why she believed in matriarchy. The second came from a Lacanian, almost in tears, passionate at her stupidity. The third told her that she had a choice – empirical research or speculation. But not both. Not these two things at the same time.

Irigaray's suggestion was that women's 'relational' identities provided a good model for making relationships between two different subjects, and for living together. She suggested that the architecture of our homes should provide the two sexes with separate spaces as well as a place to invite the other.

For Irigaray, the potential of the insertion of that tiny preposition, the word 'to', into the phrase 'I love you' making 'I love to you' suggests a new social order of relations between two different sexes. Language provides an utopian impulse. Prepositions possess a strong suggestive role. Capable of changing everything around them, they provide a means of making connections between two.

Philosopher Michel Serres describes prepositions as *angelic*. The 'angel' described by Irigaray is one who moves between, thwarts all repetition and challenges existing boundaries.

Design education encourages us to be specific and definitive: to behave as if architecture is an object separate from us. Rather than a place we can have a relationship with. How would architecture be if we considered it in terms of prepositions, in terms of the transformative role prepositions play in making relationships between people, and between people, things and places?

Intervention 4:
closer
Site:
toilets on the ground floor
Action:
to read in candle-lit toilet cubicle with the door locked and my voice transmitted via loudspeaker into the other cubicles where the audience were located
Words:
08.05.00:
New York Earth Room
7000 oaks
31.05.00:
Touch
A long coastal walk
Between the devil and the deep blue sea

26.06.00:
The Vertical Earth Kilometer
26.08.00:
Exit: come on in
09.11.00
It's quicker by rail
Holidays afloat
Rounding the Cape of Good Hope

I've recently got close to an artist. In the days and months as we moved towards each other, he sent me a series of postcards. Some were of landscapes, others of land art. All are reference points of our mutual topography – a mapping of the merging of our emotional, creative, intellectual worlds.

Before this transformation, I wrote about his work, as an independent critic (or so I thought). Recently, he told me that he found it hard to recognise his work in my comments. What are the reasons for this lack of recognition? He was sceptical about the ability of someone in love with him to be objective. But in my opinion, the critic can never be objective, there is always something about the self at stake...

For me, the critic is a travel writer, always going far from home, invited as a guest into someone else's place. To enter another's space necessitates movement out of one's own territory, it involves trust on both parts. To engage with another is to journey from what is already known towards what is as yet unknown. To encounter another requires a willingness to connect, but also to let go, to take risks. Some critics travel like tourists, crossing vast territories but remaining unchanged. Others, like me, are constantly pulled out of the familiar toward the strange, impelled by a desire for transformation. This is nothing if not subjective – a total immersion in the other – in order to return anew to the self.

What is criticism? Who can be a critic? Does the distance between the critic and her subject matter make a difference? If the critical viewpoint is meant to be one of 'judgement', is this judgement subjective or objective? Being objective seems to imply that I perceive a distinction between myself and the thing I am critiquing – that I occupy a position of distance and

How to take place (but only for so long)

disengagement. To be subjective is to get so engaged in the other that one gets close, close enough to get lost, close enough that intimate proximity becomes a problem rather than a delight.

Intervention 5:
telling it as it is
Site:
the courtyard
Action:
to walk around the outside courtyard reading the words that other participants had chalked up on the wall, words which describe the roles the participants identified with
Words:
Traveller
Tourist
Migrant
Exile
Refugee
Nomad
Architect
Historian
Teacher
Theorist
Critic
Researcher
Writer
Storyteller

My early childhood made me into a traveller. For years I travelled physically all the time. Recently I've been moving frequently. But I've not been travelling in the same way. My body stays still. Through writing, reading, teaching, researching, I lose myself in other people's heads. Sitting still with students coming through the door in an endless stream, I am on the move.

We all like to tell stories. Women like to tell stories, stories about people they know. These stories are called 'gossip'. Men tell stories too. Katerina Ruedi once described them as 'anecdotal teaching', old men taking up time telling irrelevant stories about themselves. How are these stories any different from those that women tell?

Perhaps it is because the stories women tell are often about journeys. They describe where they have come from, where they are going and what it is along the way. The feminist adage, 'the personal is political', favours the travelogue or autobiographical journey.

At a talk a few years back, Susan Rubin Sulieman described the often painful process of re-reading parts of her older work, the way she would cringe at some of the stories she has told, stories that with hindsight she felt did not resonate, stories that were not a microcosm of a greater whole, not a detail of a larger pattern, just personal outpourings. bell hooks has expressed a worry about our need to confess. Why do we tell stories? Is it to confess?

When I tell stories to my students, I do so to try and shift the power dynamic between us, to make a space of trust. By telling them my stories of difficulty, 'I've been there too', I also reveal weaknesses, aspects of myself that disempower me. To travel this distance I put myself at risk.

For me, teaching is an encounter with another. As such, it can only ever be the taking of risks, the refusal of certainties, the acknowledgment of not-knowing. My students want answers and I won't give them. Even if I have them. I love the questions.

Epilogue:
spinning
Site:
middle of the courtyard
Action:
to spin and speak (without falling over)
Words:
...

Irigaray notes that when her mother goes away, the little girl does not do the same things as the little boy. She does not play with a string and a reel that symbolise her mother. Because she and her mother are of the same sex, her mother cannot have the object status of a 'reel'. Instead the little girl is distressed. She plays with dolls – a different kind of object from the reel. She dances, 'this dance is also a way for the girl to create a territory of her own in relation to her mother'. In her dance, she spins around, de-stabilising existing connections between herself and her place, making new ones between herself and her (m)other. She creates 'a vital subjective space open to the cosmic maternal world, to the gods, to the present other'.[10]

Irigaray's daughter spinning to make room between her and her mother resonates strongly for me. I imagine being 5 again, spinning round and around in the middle of a room. Only stopping when the furniture, walls and floor begin to revolve around me, when everything slips out of place...

I have found those words of Irigaray so inspirational, for so long, that is hard now to consider them critically. I have come to take for granted that for me the state of 'de-stabilisation', my slipping of 'out of place', is a positive place to be. Why? Because I have connected letting go with liberation and assumed that my spinning is transformational, that by turning I can create future possibility, spin a new place, move no place into a new place to be. But the giddiness and the freedom of letting go that spinning offers can only last for so long. I, for one, cannot maintain the sense of disorientation that the place of spinning produces. As I suggested at the beginning of this piece, the Paris conference got me thinking differently about myself and my work, about my then almost uncritical belief in the utopian horizon of French feminist theory. But I know now there are difficulties with giving and with the place spinning makes. Instead I realise that I need to take, to take place, for myself, if only for so long.

Taking place

How to take place?
Who to take place from?
How to give place back?
How to take place? (but only for so long)?

---

[10] Luce Irigaray, *Je, Tu, Nous: Towards a Culture of Difference*, London: Routledge, 1993, p.59.

Chapter 6

Jacques Derrida
Charles Bernstein
Ann Oakley

Katie Lloyd Thomas

Robin Evans

Catherine Ingraham
Brigid McLeer

Andrea Kahn
Stan Allen

BUILDING WH
BEING IN IT:
notes on drawing 'otherhow'

Hélène Cixous

Rachel Blau DuPlessis

Chapter 6   Katie Lloyd Thomas

Julia Brown

www.inplaceofthepage.co.uk

Christine Hawley

Julia Kristeva; Karen A. Franck

ILE

Ron Silliman

Cris Cheek

Sue Thomas; Sergio Bessa

BUILDING WHILE BEING IN IT: notes on drawing 'otherhow'

Lately, I have been making architectural drawings again. It is no easier for me than it has ever been. To draw, to drag, to extrude, to pull the monotonous lines out across the page. To make the horizontal line meet the vertical line and leave no evidence of the very different arm movements that produce them. To apply an even force to the whole length of the line without it petering away to nothing, or the nib bending from too much pressure. To draw in ink on trace is to try to omit tone, colour, weight, gesture, remains. But my tracing paper is always scarred, torn, scraped, marked – a troubled surface.

Catherine Ingraham has called the architectural drawing a *lament*.[1] As the infant's cry is an attempt to bring the absent mother into presence, so too the architect's marks stand in for her own lost object – the distant building that she will never shape. Perhaps so for some architects, but for me these reluctant lines have only ever been an incomprehensible obstacle to the building. If they lament a lost object, it is only that other drawing, the one that we draw, says Hélène Cixous:

When we were little. Before the violent divorce between Good and Evil. All was mingled then, and no mistakes. Only desire, trial, and error.[2]

In that other drawing: the tool, the page, the thing drawn, the body, the self attending to it and the passing of time, all folded into a continuous surface. In 'that other drawing' there was no outside, only a collapsing of edges and a drawing together. But the architectural drawing is more than difficult, for it can never, no matter how many refinements and re-tracings I make, be anywhere but at arm's length, well and truly outside me.

I have been making these recent architectural drawings voluntarily (for the first time neither for a salary nor towards a qualification) because an artist, Brigid McLeer, who knew my theoretical work on architectural drawing asked me to translate a series of her graphic 'textplans' into drawings of a building. Over the years I had produced a number of texts

---

1 Catherine Ingraham, 'Losing It in Architecture', in Francesca Hughes (ed.), *The Architect: Reconstructing Her Practice,* Cambridge, MA: MIT Press, 1996, p.10.

2 Hélène Cixous, *Stigmata: Escaping Texts,* London: Routledge, 1998, p.21. The shift in tense from 'we draw' to 'when we were little' follows Cixous' text.

critiquing architectural orthography from a feminist perspective in an attempt, perhaps, to place my own alienation. Each of these set out subversions and disruptions that could provide escape routes.[3] But, finally, given a chance to explore drawing in an art context, free from the practical constraints of my professional discipline, I did not utilise any of these possibilities. Brigid did not tell me to work according to the orthographic conventions of my discipline, and did not prescribe any specific output. But nevertheless, I chose to set myself the rules of my discipline and adhered to them absolutely.

The lines of the final drawings I produce for *In Place of the Page* are those of the conventional architectural drawing and although the building they describe consists of soft paper walls, fantastical steam gardens, strange layered doors and a bridge structured out of letters; people are often convinced by them and ask us 'Are you going to build it?' No rules have been broken, no traditions ignored. There is nothing in the codes of the drawings that marks them apart from any others and I struggle with their execution as I ever did. There is no return to 'that other drawing', the absorbing self-defining activity of the years before my architectural education that is now rarely pursued, except sometimes in front of the mirror, over and over, until the face, charcoal, hand, eyes, marks find me and lose me all at once.

Instead I *perform* the process of the architectural drawing. I misapply its logics, read it many ways not one, let it produce strange monsters, use it to force the disparate to cohere. My delight is in what the conventions can do; how they can make something out of nothing, make paths from one image to another where none would seem to exist; how they insist on repetitions, can't help but produce floors, walls, openings, roofs. What was a critical engagement with the architectural drawing from a clear feminist position – a concern with its omissions, repressions, hidden motivations and operations –

---

3 Among my different papers dealing with issues of architectural drawing are: Katie Lloyd Thomas, 'Conceiving Architecture Through Colour: A Feminist Enquiry into Architectural Representation', *Visual Culture in Britain*, 2 (1) (2001): 17-34; 'Crossing into the Line', paper given at the *Alterities* conference, June 1999 and published in the proceedings of the *International Seminar on Gender in Arts and Culture*, Slovakia, October 2000; 'Lines in Practice: Thinking Architectural Representation through Feminist Critiques of Geometry', *Geography Research Forum*, 21 (2001): 57-76.

has become a practice of drawing in which, without identifying with it, I discover what its potentialities are in my own hands.

In this piece, I will explore this shift in relation to the drawing processes I use in *In Place of the Page* and some of the ideas I have explored in previous critical analysis. My example is symptomatic of broader shifts within discourse on architectural representation and also within feminist work. Since Robin Evans wrote his seminal essay *Translations from Drawing to Building*,[4] in 1984, it has become common to acknowledge that the architectural drawing is not simply a reductive and failed representation (or 'pre-presentation' since the drawing is usually prior to its object) of a building-to-be. Rather, the drawing is understood as an operator, allowing what Stan Allen has called 'a transaction' to take place between the abstract realm of geometry and the material stuff of building.[5] As such, lines and geometries do not have to be considered prescriptive or limiting – because in transacting with a diversity of conditions they will produce different outcomes. My changed relation to the architectural drawing follows this trajectory.

Within feminist work, at least in the dominant traditions in the UK, the USA and France, there has also been what the conference *Alterities* identified as a move from 'feminist politics' to 'feminine poetics', which is mirrored in the microcosm of my own work on the drawing. While earlier, feminism formulated alternative histories and ways of knowing, developed critiques of social conditions and new constructions of subjectivity, a variety of practices are now emerging which take a critical position in relation to mainstream disciplines, and develop new ways of working which offer potential for feminine practice even if they are not defined as 'feminist'. Examples might include Cixous' *'écriture féminine'*, Tracey Emin's art work or the urban work of *muf art and architecture*. Such practices cannot be understood precisely through the terms of feminist critiques and always

4   Robin Evans, *Translations from Drawing to Building and Other Essays*, London: Architectural Association Publications, 1997, pp. 153-94.

5   Stan Allen, *Practice: Architecture, Technique and Representation*. Amsterdam: G+B Arts International, 2000, p.6.

exceed them. They slip away, refuse to play by the rules, and sometimes tempt us to make overstated claims for them. But this does not mean that we should remain within the more safely defended feminist activities – telling our stories, voicing our experiences, even establishing our identities – even if when we move away we can no longer say for certain that what we do is feminist.

In what follows I explore this mis-mapping between critique and practice. I suggest that critique provides a place outside, from which one can return to practice and to 'building while being in it'.

Brigid McLeer began the project *In Place of the Page* in 2000 with the aim of engaging with notions of space and inhabitation:

All kinds of spaces – and our abilities to inhabit them – will be explored; domestic, social, architectural, transitional, narrative, textual, imaginary, bodily, virtual etc. etc., with the main questions being, in what way do these spaces/places make room for, delimit or interact/interfere with senses of identity?[6]

The project has had three main phases, each of which takes place in a particular medium with its own spatial limits and possibilities. Initially ten participants addressed Brigid's questions to each other through a six-month email correspondence. The emails frequently explored the places in which they wrote to each other and the particularities of the virtual space in which their exchanges unfolded and the nature of the location which their exchanges constructed:

I am increasingly fascinated by the architextures here – the windows, the browsers, the navigational equipment, the search engines, the links, the gateways.[7]

Sitting under the apple tree at the bottom of my garden with my laptop on my knee and a glass of apple juice beside me.[8]

---

[6] Brigid McLeer, *In Place of the Page Postcard No. 1*, February 2000.

[7] Cris Cheek, email 20 March 2000. All emails cited here can be accessed at *www.inplaceofthepage.co.uk*

[8] Sue Thomas, email 15 May 2000.

Building while being in it: notes on drawing 'otherhow'

Using these texts as material, Brigid has been making what she calls 'textplans', graphic samplings constructed in digital space, and subject to the tools of the program she uses. In the textplans, passages of text may be removed from their original context of meaning, clipped, cropped, doubled, rearranged, laid over each other in thick tangled weaves or single letters, and may be expanded to become a form or field in their own right. Brigid also imports other drawings – her own sketches scanned in, or the print distorted and enlarged to reveal its grainy materiality, or blocks of colour, shapes and signs which are part of Photoshop's graphic inventory. It is these textplans which Brigid hands to me for reinterpretation.

Signings and de-signings: voices in the space of the page
I found the entrance first. A bridge across a watery shifting place to a series of doors where one could deposit spoken and written words... Will the exit be the last page to arrive?

The first drawing I made, the bridge (to this dialogue) is in part constructed out of some lines and the letters 'b' and 'r'. Other letters, 'i' 'd' 'g' 'e' disappear in my tracing from the textplan; they are not useful and are edited out. This word has, in turn, been selected by Brigid from one sentence in a lengthy email dated 31st March 2000:

In the 'introductions room' I placed a caravan which is really a bridge – well, I'm still in that place but far nearer the other side than I was – moving towards that phrase in my text plan – home, an edit, a decision.

and signed 'Kirsten Lavers'. Kirsten's email which included a quotation from Shane (16 March 2000) produced another textplan, a rectangle that I read as a door at the end of the bridge where gift-words could be posted, and a third textplan, which determined the plan of the bridge at a much larger scale, was a graphic response to an email from Sergio (1 May 2000).

There are 56 emails from the eleven contributors to the first phase of *In Place of the Page* and each is signed in the conventional way:

Brigid / B

jane / Jane

Sergio / S.

cris

s / sue / Sue

Kevin

Julie

Gregg

Patricia P. Potter / Pat / Pat Potter – Alabama Studio

Shane

Kirsten Lavers

> Fri. 31st March
>
> my book would be Paul Auster's - The Invention Of Solitude - a slim book in two parts - quite, quite different from each other - I'd like now to quote a section in which he talks of coincidences but I can't since I'm on a train from Norwich to Cambridge I'll post it later, but for now other coincidences are emerging in my In The Place Of The Page mailbox .....
>
> Brigid sent me a text plan containing the words - home a edit, a decision Patricia and Jane have both mentioned current engagement with making new living/working spaces well, just to register with you all that I'm there too and am moving towards a space with no doors,furnished window sills, no curtains and pipes that do not attempt to conceal themselves am aware that this dialogue already seems to have visited three rooms at the very least - wondering are there doors to these rooms? indeed to this dialogue? do you mind if I open them, remove them? refuse to close them?
>
> in the "introductions room" I placed a caravan which is really bridge - well i'm still in that place but far nearer the other side than I was - moving towards that phrase in my text plan - home an edit, a decision and then Shane's description of the book "she's" living in arrived in my mail box it strangely echoes my own recent experiences of living in a book - or more accurately - books, an experience I share with cris (hope you don't mind me mentioning it here, cris, but it seems pertinent) - we're collaborating (maybe that should be in the introductions room) on various projects (we call them occurrences) all flowing in some way from things of our own and things given to us that in some way could be argued as not worth keeping
>
> we've constructed for ourselves two spaces which we've recently joined into one, spaces in which we've imagined that we're being held hostage, we have a deadline - by May 2001 we have to crack the code of this space/s or ?........... the spaces are two ring bound hard cover books. One book contains 100 pages all numbered 99/100 taken from 100 books spanning a century of publishing 1897 - 1997 and all argued by us and others as not worth keeping - the other book contains 101 pages all numbered 101/102 from the same books plus one! these pages are the only remaining easily legible remnants of their host books which have all been shredded - only very recently cris and I have been weaving these shreds at the same time as bringing these book/spaces together and our making first shared attempt at the task of decoding/writing/reading them.
>
> As you can perhaps imagine these spaces are bizarre, fascinating and disorientating - a mix of paper textures, formats, fonts, illustrations, genres make the journey through their double page spreads not dissimilar to walking across the flat plains of East Anglia where we both live - each turn of a page (each step) subtley expand and alters the terrain invariably much of the territory remains familiar in the frame around the page (especially if its small - like the Paris map pages or Ruskin on War) - each turning reveals only what it has concealed, concealingi what was revealed. And then occassionally there is the dramatic turning of the page from the Guiness Book of Records or The Bride's Book or Greenhouse Gardening when, like suddenly reaching the top of a hill or breaking out of a thick forest everything suddenly changes before you and familiarity can only be found be turning around and revisting where you've already come from.
>
> it's a fascinating process, I do feel as if I've been living in these books, they are old friends now that still have the potential to excite, offer gifts, inspire, confuse and surprise me, we've each made an individual walk through their pages - documented now as first and second readings - we're no nearer breaking the code - the spaces are now joined, the partition wall has been dismantled they can only be read together and by two people - our next walk will take place on 29th/30th April over 24hrs - we'll keep you posted if you're interested (sorry about the mixing of metaphors!)
>
> here's the excerpt from Shane's mail which really echoed for me -
> According to custom, the occupants (now her lifelong friends) give her gifts - words - which she uses to construct her own houses. These gifts become interwoven like materials for a nest - she relocates them, places their words in different sequences, thereby making her own original creation. No cheap reproductions here. But it is difficult to allow people into the conversation when it is so very intimate, almost in code.
>
> is this simply a coincidence? finally, if the above has made you curious another 'thing' to furnish the introductions room http://www.thingsnotworthkeeping.com
>
> am feeling excited and supported by what's happening here
>
> more later
>
> Kirsten Lavers

When the emails were exhibited at Nottingham Central Library in November 2003, hung as 56 bound books on the wall, each email was the front cover to one of the books. By giving equal importance to each contribution, the project appeared to be collaborative, a way of working which is a common feminist approach. In the second phase, however, Brigid takes authorship and the contexts and intentions of the writers' words disappear. As Brigid describes it:

> This process is not only a formal interpretation of the emails, it is also a kind of reclaiming of 'authorial' control, filtering the many voices through just one, my own. Recently someone said to me that they found the textplans 'muted' the original emails, and I think that is probably true.[9]

In just one textplan the author's name remains, 'more on this soon, Jane' – but otherwise Brigid deletes the names, and I have no idea whose email instigated the graphic I work with. Unlike a handwritten letter, an email has no voice, so sign of a familiar hand, it only conveys words, punctuation and some degree of spacing and there is little control of how it will appear on the recipient's screen. Brigid's textplans, on the other hand, are full of voice although they are produced digitally: colours, freehand lines, her complex spacings and overlays and each is immediately visually distinguished from the next, but the voice is no longer the contributor's, but hers.

9   Brigid McLeer, unpublished paper.

The textplans are responses to just one email (sometimes with quotations from another embedded within them), but I always need a number of drawings to construct any architectural element (in order to see it in at least two dimensions, and often to understand it at different scales). The bridge then is a multi-authored bridge, where the voices are lost. A bridge made with the letters of Brigid's name, that is the subject of the Flash movie she traced from my drawing and animated, a bridge that she now paints onto a blackboard and chalks over wildly, covers with other drawings, her own made from texts people give her, documented, deleted, covered over again.[10]

To what extent do the conventions of text and its mode of production construct or cover over the author/s? Feminist artist Mira Schor explores the differences between the standardised handwriting we are taught to copy as children, and the development of our own adult hand in her painting 'Personal Writing'. Painted on a large scale, each letter 12 inches wide, the deviations between the two suggest the personal inflections which are possible within an order determined by power relations and a (patriarchal) order. Within a machine-generated text, such as the emails, the letters become completely standardised. For American L=A=N=G=A=G=E poet Ron Silliman, the increasing tendency of text to lose all evidence of its human production is symptomatic of late capitalism. He sees the contemporary word in terms of Marx's commodity fetish:

Torn from any tangible connection to their human makers, (words) appear instead as independent objects active in a universe of similar entities, a universe prior to, and outside, any agency by a perceiving subject.[11]

In *Glas*, Derrida reads Hegel's (desiring) voice into his philosophical texts which are supposed to be objective revelations of truths beyond their author. He suggests that Hegel signs his text despite this: he addresses the reader directly and thus

---

10 The Flash movie can be viewed on *www.inplaceofthepage.co.uk*. Brigid produced the chalk drawings at a group show *Watch this Space*, Phoenix Gallery, Brighton, 24 April–29 May 2004.

11 Ron Silliman, *The New Sentence*, New York: Roof Books. 1995, p.8.

reveals his desire to be understood by another and he uses the metaphor of the eagle/'aigle'/Hegel, soaring above the earth and looking down.[12]

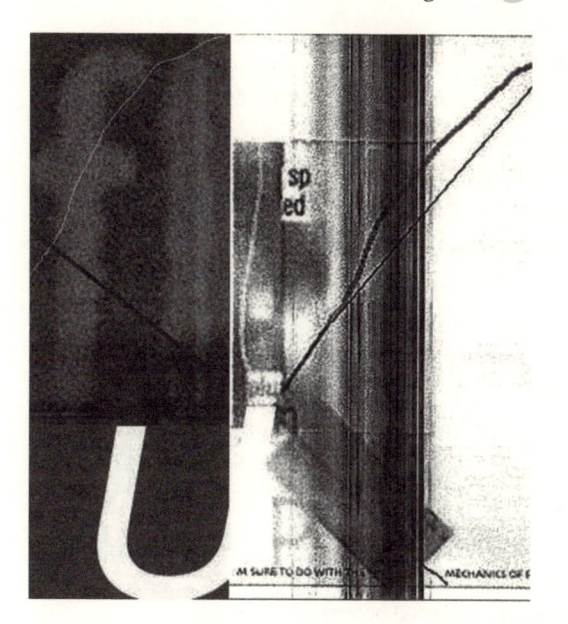

Brigid's textplans incorporate both the gestural (her line drawings, sometimes her handwriting) and the standardised (typefaces, symbols, digitals functions from the graphic program) and sometimes the evidence of their making (misprinted photocopies, the edges of printed words which have been cut and pasted) but they turn the authored emails into materials for her own use. This is not a transformation which tries to preserve its multi-authorship; rather the page and its conventions are shown to be a space in which voices can be lost, meanings can be spliced, decontextualised and words can be used to build all kinds of constructions.

For the architectural drawing, the conditions are different. In a false etymology, to de-sign becomes the erasure of individual/s, the making of an object which appears not to have been (co-)authored. Architecture, unlike text, is almost always produced by a number of people, by a team of architects in

[12] Jacques Derrida, *Glas*, trans. John P. Leavey, Jr. and Richard Rand, Lincoln, NE: University of Nebraska Press, 1986.

consultation with many other specialists. But, like the email, the orthographic drawing is particularly 'voiceless'. It appears to have been produced by a single hand, although the title block may record the initials of those who worked on it, or labels in another handwriting might be added, or a story of amendments and conversations might overlay the drawing in markers and hasty gestures.

To draw according to architectural conventions is to de-sign, to resist personal expression and favour the realm of universal truths of geometry. Although it is often said that architectural drawing allows communication between the architect and the builder, historically it produced a separation. The widespread use of the orthographic drawing really begins with the founding of the architectural profession as we know it, just after the French Revolution when the first 'school', the Ecole Polytechnique, was set up to train military engineers and surveyors. Gaspard Monge developed 'descriptive geometry', the comprehensive system of drawing which could define the building in its entirety and is still in use today. Whereas Renaissance architects had used only the 'footprint' plan and a few elevations, the number of drawings produced by architects post-Monge increased exponentially (and in Britain the new technique coincided with a significant decrease in paper tax). The new technique allowed the designer to be completely separated from the messy process of building – and encouraged a focus on form over the physical and knowing crafting of the parts. The standardisation of architectural drawing, its de-signing, coincides then with the emergence of the profession. The producer of the drawing is singular, ungendered and invisible.

Enfleshings: the word is (the line is) material

```
Mostly I trace Brigid's drawings. Tracing is a kind of
copying, always from something already there. Tracing,
unlike printing or rubbing or photography, prefers
outlines – it is 'outline loving'. It can transform the
amorphous into the prescribable. Some changes in colour
or tone, some densities of texture and information, are
mopped up by the thick greyness of the tracing paper,
```

```
soaked into it, blotted, lost to the copy.
ffffffffffffffffffff......fllllllll......fl... fl... fl... flui..
the mechanics of
fluids
```

'f' enlarged on the photocopier, out of focus, fading into the background, enfeebled, enfleshed, ephemeral. This 'f' commands the reading eye's first attention on the right hand side of three of the textplans, and varies in colour within a spectrum of watery blue greens in each. The first plan I looked at is grainy and textured and includes the words 'hold in' almost absorbed in some streaks of lighter blue below a hovering deep coloured 'fl'. Hand drawings, particles of pixelated ink and a band of lines from a misprint on the copier are visible in the image.

Tracing over this plan, I found a detailed section through a 'word-wall'; a threshold through which one speaks without being able to see. The voice is amplified through a series of chambers to become vibrations on the other side, and the breath is condensed in a flat glass box to become visible. It is not possible to translate the ambiguities and material qualities of the textplan in the lines of an architectural section. Instead, my response to them seems to be a concern with ephemeral elements usually not considered as architectural materials. Rather than the usual 'softwood frame', 'rockwool insulation' or 'hardwood cill', I label the drawing with 'cold air', 'light', 'whisper wool', 'condensation/glass/breath', and 'reverberate/ echo/sound'.

I've only recently read the emails from which this textplan was constructed: Jane Rendell's email on 16th March 2000 and Brigid's response to it on the same day, and they are also concerned with thresholds. Jane identifies 'a real anxiety about threshold spaces' in architectural manuals with their devotion to details of window cills and doors and their particular attention to membranes and protection. She refers to 'The Mechanics of Fluids' which she says 'don't fit happily with solid constructions keeping the water out'. The two emails explore many slippery figures such as 'the fluid', 'the unstable', 'the moving', and 'spaces between', figures derived in part from discourses on tactical feminine subjectivities: Rosi Braidotti's nomadic subject, or Luce Irigary's figures of the relational between – air, water, breath and mucus.

The architectural threshold tends to withstand or at least control movement across it and the architectural drawing, with its continuous unbroken lines (doubled to signify a wall as if to reinforce its defensive status) also pre-empts such a securing of boundaries. It cannot represent the porous, the fluid, the almost immaterial, as Christine Hawley has described:

How does (the line) deal with something that is reflected or immaterial, a shadow? How does it deal with that which hovers, threatens, glides or melts?[13]

Within the logic and conventions of the architectural drawing whatever is non-linear is edited out. The blurred fields of graphite which occur on the surface of a hand drawing as it is worked or, equally accidentally, through the processes of reprographics, are described beautifully by Justine Clark in her essay, 'Smudges, Smears and Adventitious Marks'.[14] They are always only the 'underside' of the architectural drawing revealing as she says 'the materiality of the drawing' but remaining in themselves overlooked and unread.

The architectural line withholds its own materiality, and withstands seepage and crossing into. In this, it has much in common with the letter or word, through which we look to see the meaning beyond. Through techniques of interference – changes of scale, sampling and decontextualising, distortion, overlays – Brigid reveals the material of written language and refuses to let her work be read as its linguistic content alone. Silliman suggests that textual processes such as these which bring the signifier back into view can provide 'experiences of that dialectical consciousness in which subject and object, self and other, individual and group, unite'[15] – over-emphatic

[13] Christine Hawley, 'Invisible Lines', in Francesca Hughes (ed.) *The Architect: Reconstructing Her Practice*, p.169.

[14] http://www.architecture.auckland.ac.nz/publications/interstices/i4/THEHTML/papers/clark (accessed 30 April 2004)

[15] R. Silliman, *The New Sentence*, p.17.

claims perhaps for the political effects of a critical writing practice. Poet Charles Bernstein is more circumspect. Terms such as 'materiality' used in relation to poetry are 'like the personal pronouns, shifters, dependent for their meaning in the particular context in which they are used',[16] in this case presumably the context of avant-garde writing practices in the twentieth century. Nevertheless he shares Silliman's concern with the visibility of language:

The tenacity of / writing's thickness, like the body's / flesh, is / ineradicable, yet mortal. It is / the intrusion / of words into the visible / that marks / writing's own absorption in the world.[17]

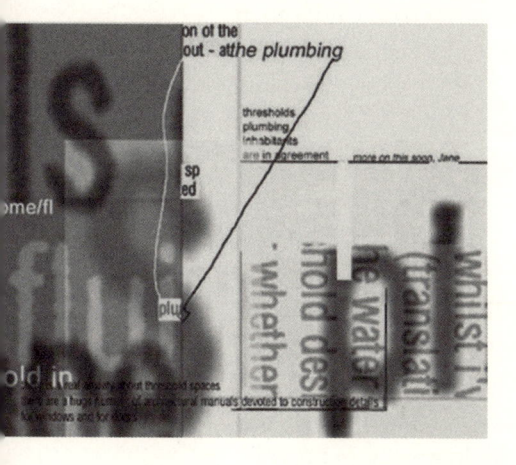

Initially my research into the architectural line was also concerned with revealing the fleshy substance of the drawing. I developed a critique that grew from Catherine Ingraham's claims that, in denying its own materiality, the geometric line represses 'the counter-geometric life, the everyday life of the inhabiting body'[18] and 'is the trace of a soon-to-be dead (female?) body, a sexuality that will be built over, buried and deeply lamented'.[19] This body – abjected and erased from the line – I took to be Julia Kristeva's maternal body, the body which threatens our stability as subjects and is conflated with

16 Charles Bernstein, *A Poetics*, Cambridge, MA: Harvard University Press, 1992, p.157.
17 Ibid., p.87.

18 Catherine Ingraham, *Architecture and the Burdens of Linearity*, New Haven, CT: Yale University Press, 1998, p.110.
19 Ibid., p.112.

the body of woman who is also abjected by society. Kristeva develops these ideas in her work on painting, understanding the geometric line within the terms of the symbolic, and colour within the terms of the destabilising semiotic. I looked to artists such as Helen Frankenthaler who, according to one critic, 'takes colour into line and enclosing shape, creating imagery that is between or both line and shape',[20] and at the painted-over loose ink photocopy drawings of Bracha Lichtenberg Ettinger. I analysed Will Alsop's use of colourful gestural paintings during the design process. The success of his paintings in encouraging open-ended dialogue about the designs could, in part, be explained in Kristeva's terms. She suggests that Giotto's use of colour avoids the 'identification of objects' and instead triggers 'the instinctual and signifying resources of the speaking subject'. In other words, the painting 'reaches completion *within the viewer*'.[21] Perhaps, I argued, Alsop's paintings also engage the viewer in an active process of constructing meanings.

But it now seems to me that a concern with the 'flesh' of writing or painting can only be transposed to an architectural practice in a limited sense. Only if the architectural drawing is considered a representation, can the line be a signifier, and this is an inadequate definition of a notation whose task is also to aid in the construction of a future artefact. As Allen has written, 'The architectural drawing is *transitive* in nature, uniquely capable of producing something new from something else.'[22] At least, for the architect and the builder, it is not a referent, but 'a set of instructions for realizing another artifact'.[23] This is not to deny that, for the most part, architectural drawings *are* read as representations, and that at least since the Renaissance when they became widely used, the lines, angles and secure thresholds found within them are replicated in the buildings that they generate. The orthographic technique with all its omissions and tendencies leaves its mark on the objects it produces, but it is not simply a failed

---

20 Julia Brown, 'In Pursuit of Beauty: Notes on the Early Paintings of Helen Frankenthaler', in *After Mountains and Sea: Frankenthaler 1956-1959*, New York: Guggenheim Museum Publications, 1998, p.50.

21 Julia Kristeva, *Desire in Language: A Semiotic Approach to Literature and Art*, Oxford: Basil Blackwell Ltd, 1982, p.231.
22 Stan Allen, *Practice*, p.36.
23 Ibid., p.32.

or reductive representation. It may make only limited instructions (make this a certain distance from that, in a certain relation to another, above, below, within) but there are an infinite number of contexts and materials in which it can be put to work and as many outcomes.

The thin ink tracings I make of Brigid's watery textplans cannot record their grain and leakages, but they are able to describe an object which could record the slight physical manifestations of speech, that will allow some seepage through its threshold, that will use breath and sound as materials. This is not the fleshy and materialised drawing process that I imagined when I formulated my 'feminist critiques', it barely inflects conventions, but it produces an architecture of material, remains, condensation and fluids. It is an 'other' drawing practice, not because of the nature of the objects (drawings) it produces, but through the peculiar (fluid) processes of its unfolding.

Mobilising the page: non-linear strategies in writing and architectural drawing

Rules for interpretation of textplans:
1. attribute to each drawing a conventional scale: 1:5, 1:20, 1:50, 1:100, 1:200, 1:500
2. read each drawing as a standard orthographic projection: plan section elevation
   (later I decided to include sketch not to scale)
3. each drawing may be interpreted at more than one scale, in reverse, and in repetition
4. interpret words as labels (of sorts) referring to: descriptions titles activities materials

It was only by setting these rules that I was able to look at these drawings as an architect.

One of Brigid's most poem-like textplans is constructed almost entirely of letters with the addition of five sets of bisecting lines and one erased line. In the centre of this knot is what appears as a hole (or a birthmark?) torn in the page allowing a glimpse through to the original text from which the letters are culled – these are white letters on a dark grey, a reminder perhaps that they were first read on a computer screen. Some letters are sampled and reassembled but one of the sentences is retyped in full:

If a place is a set of relationships then those relationships can be made of anything.[24]

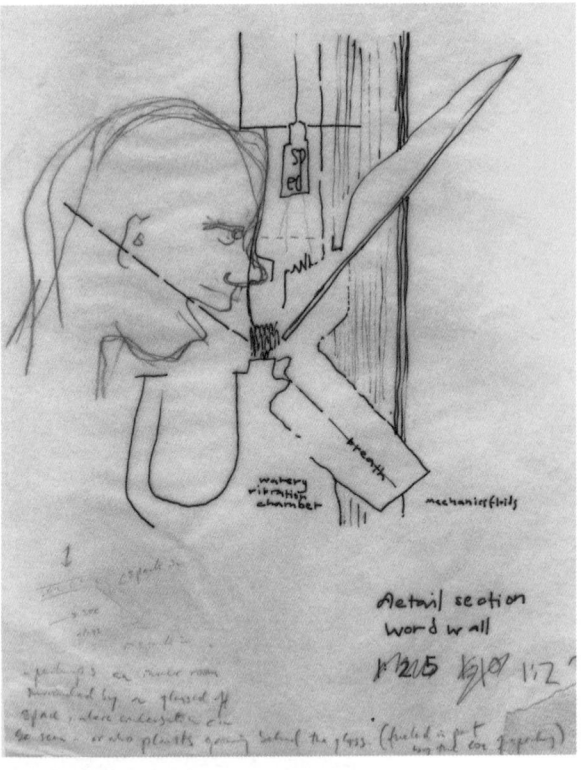

The construction of the textplan embodies these words, it is 'made of anything' – of letters and assembled to form 'a set of relationships', and it sets up the page as a 'place' which the reader navigates.

The spatialisation of the text invites the active participation of the reader, and avoids the way that 'books just pull you straight from page one to the final page… like a river'.[25] Since Mallarmé scattered the parts of his poem 'Un coup de dés jamais n'abolira les hasard' ('A throw of the dice will never

---

24 Brigid McLeer, email, 9 May 2000.   25 Sergio Bessa, email, 1 May 2000.

abolish chance') across the page,[26] avant-garde writers have been challenging the conventional linearity of text. For feminist poets such as Rachel Blau DuPlessis, this tactic helps keep the work open, avoiding notions of a 'proper' reading, and the singular authorial intention. As Brigid describes it:

The reader enters into the work... as such they are... newly responsible for, and often surprised by, the meanings that they, in collaboration with the author make. The page then is a metaphor for a complex and lived, negotiable space.[27]

The architectural drawing does not propose a direction of reading in the same way, and is in fact usually read in relation to a number of drawings. A sense of the whole only emerges from reading between the different scales and views – and not usually in any particular order.[28] When I trace over this textplan, the words are erased and the resulting formal arrangement of lines loses its radicality. The architectural drawing is already spatial in the sense of its relationship with the page, but nevertheless has its own linearities. Ingraham has identified at least three: (1) the literal preference for the straight geometric line; (2) the assumption of a linear translation from drawing to building; and (3) the line between two binary terms such as line/not line.[29] One might also suggest a fourth – the 'directions' for reading an architectural drawing are given through its conventions: a limited set of projections, scales, and precise codes of what a particular line or combination of lines may mean. One can, of course, read the lines for their abstract relationships – but within the context of the discipline, their meanings are determined.

My re-tracing is subject to these conventions: rulers and stencils re-standardise the typographic curves according to another mechanism, words become the outlines of solids, a flat

---

26 The poem was first published in 1914 and written in 1897.
27 Brigid McLeer, unpublished paper.
28 Andrea Kahn has pointed out that the architectural drawing is only read singly when removed from its context. For more on the multiple and composite nature of architectural drawings, see Andrea Kahn, 'Disclosure: Approaching Architecture', *Harvard Architectural Review*, 8 (1992): 2-21.
29 See her book *Architecture and the Burdens of Linearity*, and Catherine Ingraham, 'Lines and Linearity: Problems in Architectural Theory', in Andrea Kahn (ed.), *Drawing Building Text*, New York: Princeton Architectural Press, 1991, pp. 63-84.

1. vapour
2. movement
3. vibration
4. warm air
5. cold air
6. condensation
7. whisper wool
8. fluid chamber

doors to this dialogue

edge suggests a possible ground and my drawing is identified as an elevation at 1:50. I photocopy onto acetate, reverse the image and it replicates into an easy symmetry. Both the conventions I adopt (as an architect) and the tools I use (of an architect) determine the way I read (trace) Brigid's 'open text' and the 'building' which my reading produces.

To do away with the conventions of architectural drawing would also be to do away with the discipline (how then to work to budget? how then to acquire statutory permissions?) although of course buildings can be made through other processes. Nevertheless, some architects use the drawing as a site where the usual linear unfolding of the design process can be interrupted or infiltrated. In his early drawings for the Jewish Museum in Berlin, Daniel Libeskind did not draw the lines of buildings and streets, but constructed the site through lines of relationships between people, textures and texts. As the page has made possible the translations in *In Place of the Page*, Libeskind uses it as a site of cohabitation for different registers of information outside architectural conventions, which can then begin to generate an architectural proposal because of the similarity between the marks. The British architectural practice muf work with overtly political intention and introduce non-conventional lines into their site plans, denoting not only physical boundaries but also property ownership and use. They also invent more complex kinds of drawings. In their digital map of Shoreditch, one of the layers was constructed out of the voices and stories of local residents. Dutch practice MVRDV generate graphs from statistical data and sometimes their two-dimensional forms inform the building design. Graphic similarities allow for travel between various kinds of information – music, text, statistics – and the architectural drawing. They interfere with the design process, without rendering the architectural drawing unusable in a professional context.

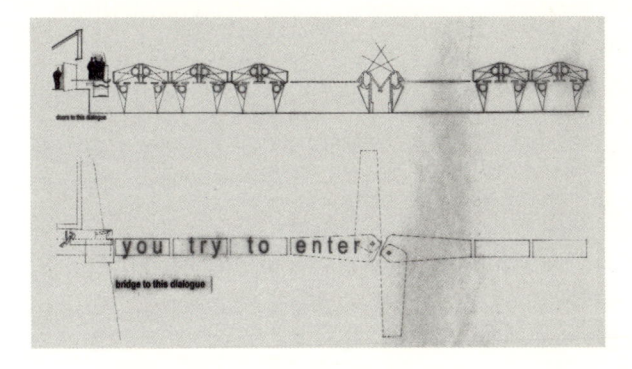

Building while being in it

In her wonderful essay 'Otherhow', Blau DuPlessis explores her own practice as a feminist poet.[30] She is careful to articulate the difference between the critical positions of modernist and postmodernist poetry in relation to tradition, and her more specific critical position as a woman for whom the tradition is also gendered. She writes:

I want to distance. To rupture. Why? In part because of the gender contexts in which these words have lived, or of which they taste.[31]

Hers is an informed response but also a visceral one, born out of a discomfort or unpleasant 'taste'. It is the taste of the architectural drawing which provoked my own more rational explorations. A bad taste is too immediate, too close to oneself

30 Rachel Blau DuPlessis, 'Otherhow', in *The Pink Guitar: Writing as Feminist Practice*, London: Routledge, 1990, pp. 140-56.
31 Ibid., p.144.

to allow for examination. It provokes only a refusal or a resistance. One needs to move away from it to begin to understand or contemplate it. This distancing, it seems to me, is part of the work of feminist critique. It enabled me to formulate coherent feminist accounts of orthography, and therefore to posit counter-drawings. Why, then, did I abandon these conclusions when it came to my own practice? In part, because inevitably I drew on other disciplines, most notably writing and the visual arts and took ideas such as the transparency of the text, the idea of a semiotic in painting and notions of linearity which do not necessarily translate neatly to architecture. And in part because these explanations are not absolutes, but instead create a temporary home from which one can begin to make judgements and take actions.

What interests me about *In Place of the Page* is not so much the objects it produces (texts, drawings, website, exhibitions) but its means of moving from stage to stage. Likewise, it is the process of making work which is emphasised in Blau DuPlessis' title 'Otherhow', that she says should also have 'caused pleasure'.[32] Although I have not changed the orthographic techniques and the kinds of drawing they make, the context of their production is radically transformed. In the first place, my drawings are always clearly dependent on the works of others. As tracings, they are literal copies of another's work, albeit heavily edited and re-imagined. At the larger scale of the project, collaborators' contributions are necessary to the growth of the project. Its progress depends on them. Even if individual voices are 'de-signed', they are central to the project, never mere marginal additions or representations. 'Otherness' – not as identity of the other, but as what they might *do* with the material – drives the project.

In the email phase of the project there are long periods of silence before a sudden flurry of exchanges. Months can go by before Brigid gives me some new textplans to work with. If one of us produces something new, it is likely to prompt the other into action. The other side of dependence is waiting. Carolyn Steedman cites Ann Oakley's association of waiting with women's roles:

[32] Ibid.

Building while being in it: notes on drawing 'otherhow'

always waiting for someone or something, in shopping queues, in antenatal clinics, in bed, for men to come home, at the school gates, by the playground swing, for birth or the growing up of children, in hope of love or freedom, waiting for the future to liberate or burden them and the past to catch up with them.[33]

> She goes on to say that 'the other side of waiting is wanting'. The waiting in *In Place of the Page* desires the response, the work of the other. It is fuelled by the desire for what the other can make, how he or she may respond. The pace of the project is leisured, it makes time for the other, and for their tantalising absence. Its duration is palpable.

'The shell' is there but I am making another kind of 'animal' live inside it. Yet a shell will always define the animal in key ways.[34]

> I work within the conventions of the architectural drawing – its 'shell' – albeit in such a way that those conventions and mechanisms are not effaced. Despite these limits, to use DuPlessis' metaphor, my new drawing process produces strange animals. It is liberating, can go anywhere, loves deflection, to be determined by factors outside its own control. The processes in *In Place of the Page* are open-ended, and can produce multiple outcomes. That we follow just one path out of all the possibilities is clear. Perhaps this openness to possibility is an aspect of feminine practice, as DuPlessis states:

Not 'otherness' as in a binary system, but 'otherhow' as the multiple possibilities of praxis.[35]

> My theoretical work suggested problematic aspects of the architectural drawing from a feminist perspective – orthography's use of an ideal line, its attention to rigid boundaries, or its power to appear objective, a-social and a-historical. A first response to this was to search out examples of drawing

---

33 Ann Oakley, *From Here to Eternity: Becoming a Mother*, London: Penguin, 1981, p.11, cited in Carolyn Steedman, *Landscape for a Good Woman*, London: Virago Press, 1986, p.22.

34 R. Blau DuPlessis, *The Pink Guitar*, p.151.
35 Ibid., p.154.

which reversed these; fleshy, material lines replaced ideal, boundaries were made porous or shifting, drawings revealed their subjectivity and partisan affiliations. But, as DuPlessis suggests, such a strategy stays within a binary construct. The 'other' drawing is defined negatively in the terms of the dominant drawings. Ways of working 'otherhow' can escape the straitjacket of being for or against, to produce unknown methods and outcomes which may even exceed them, which are then available for appropriation or analysis by feminist practice and discourse.

I have identified aspects of *In Place of the Page* which could be attributes of a feminist practice. Contingency and openness to multiple possibilities are also central to Rosi Braidotti's figuration of the 'nomadic subject'. Taking pleasure in working with others has a history as a feminine attribute, and can be appropriated without being essentialised.[36] As yet, interdependency and the desiring temporality of waiting are little more than good 'tastes' for me – the work of positioning them within feminist discourse has not yet been done.

DuPlessis' metaphor of the shell is central to this piece because it suggests that one can work 'otherhow' even within conventions and disciplines which are problematic. If we remain within the shell, we cannot see it. The work of feminist critique allows us to move outside it, although this position may not be one from which we can ourselves build. The return to the shell gives us a place from which new constructions can begin if we wear our critical frameworks lightly – a place to build while being it, a place from which a feminine poetics can emerge.[37]

---

[36] For example, Karen A. Franck lists seven qualities that 'characterise feminine or feminist ways of knowing'. These include 'an underlying connectedness to others' and 'a desire for inclusiveness'. Karen A. Franck, 'A Feminist Approach to Architecture: Acknowledging Women's Ways of Knowing', in Jane Rendell, Barbara Penner and Iain Borden (eds), *Gender, Space, Architecture*, London: Routledge, 2000, p.297.

[37] The phrase 'building while being in it' is from Brigid McLeer, email, 9th May 2000. 'Perhaps in fact I've been doing just what I've been talking about while writing this email – not "rambling" at all but being in a place that I was building while being in it.'

Chapter 7

Brigid McLeer

STRAY SODS:
eight dispositions on the 'feminine', space and

Rachel Blau DuPlessis

| Chapter 7 | Brigid McLeer |

Joan Retallack
Peggy Phelan

writing

Elizabeth Grosz

Seamus Heaney

STRAY SODS: eight dispositions on the 'feminine', space and writing   115

1
A place is being written. For now this is the place.
The page is already written, as is the place. How does she enter the page? (She is stricken, like a pose.) She enters by stepping on a 'stray sod', so she is oscillating, displaced, off the path. What is a 'stray sod'?
Later, later.[1]

e comes nto
architecture

ruck"
ke a pose

[1] The following texts are based on my original paper 'Of Surfaces and Holes', given at the *Alterities* conference, Paris, 1999, combined with a subsequent paper, 'Stray Sods', written for 'Taking Place 2' at UNL Architecture School, London, 2001.

The visual texts on each page are a series of works titled 'Culprits' done as part of a larger project, 'In Place of the Page', ongoing since 2000 and currently the basis of a collaboration with architect Katie Lloyd Thomas. For more details, see *www.inplaceofthepage.co.uk*

2

Here is a narrative of space and seeing.
She is a mother and the who that watches her, Lacan's (or
Freud's) boy-subject. He sees that her work, like an extension
of her sex, is nothing, or rather he would say (later, when he
had grown up) that her work is/was a lack of authority, or
singularity, a lack of something obviously and definitely there.

The narrative of the early stages of the subject's development in psychoanalysis (and how it tells the story of how he is and she is not) does two important things to the other mother/mother other in order to be able to include her in its analysis. It positions her, verbally and visually, in an attenuated relationship to the masculine Father. First, she is positioned outside of a discursive relationship with the subject. For even if she does speak (and of course she does), her language is already conscribed by his 'symbolic Langue'. This is particularly so because of the second motif in this narrative – which is the fact that she is seen to 'have' nothing (castrated, phallus-less) and the attendant fact that that can only happen because he sees her from a distance, a distance enough to not see. So she is always over there, the deixis of her subjectivity only ever a position in relation to his discursive centre. Her pronominal place, an emptier one than his, is only ever a shell around 'he' (he, her, here, there), he pertains, central and stable – he is a sign, she is a linguistic and discursive annotation. And she's his position of desire, her other-over-there-and-outside-where-he-is-ness figures desire for him.

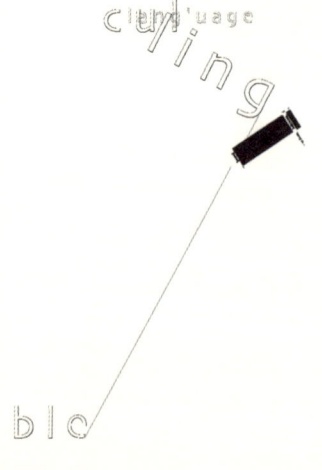

3

So she is over there in the feminine. The feminine is always 'over there', a figuration of types conceived in relation to the 'natural' masculine. She knows all this. She also knows the subjectiveness of the psychoanalytic narrative and she can speak from so many positions that feminist (and other) discourse sites for her. She can capriciously slide in and out of Kristeva's poetic, pulsing in the exchange (the choice?) of semiotic and symbolic articulations. She can slip through Irigaray's entre-ouvert, inhabiting the concavity that her proximity affords her, touch with and within her, a here that goes no further back than her. And she can occupy the annotation of her here as verbal extensions – notes and drafts, rewritings and remarkings blown against the fixity of his centre. She can speak and write on Cixous' language ladder, stepping into and onto the light, flight of detail that she writes for her.

All of these possibilities provide chances to mobilise a language that resists the ordering of masculine hegemonies, but is it also possible that the imposed spatial vacancy of the 'feminine' (in both psychoanalysis and other more social systems) might also, in its very remoteness, its relationality, its many kinds of 'gap', provide us with a dynamic kind of spatialisation, a spatial method even, that could open up the possibilities of the subject and her social and spatial exchanges.

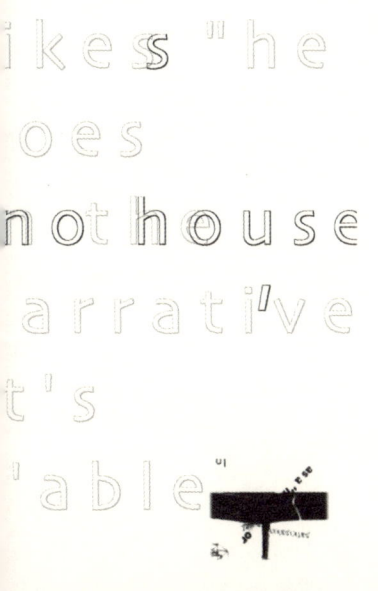

4

In an essay on 'poethical writing', Joan Retallack refers to 'feminine' textual forms or approaches.[2] In a section where Retallack redresses Judith Butler's notion of the performative subversion of parody and pastiche (in her text, *Gender Trouble*), Retallack writes 'To make real gender trouble is to make genre trouble. Not to parody but to open up explorations into forms of "unintelligibility" as feminine frontier.'

The unintelligible for Retallack is not the unintelligible of madness or hysteria, it is that which confronts reason and established conventions. It is what is inaudible or invisible to dominant thought and ideology. It is that or those who do not 'figure' in the frameworks that the dominant culture provides. And for Retallack, as for Peggy Phelan, Elizabeth Grosz and others, the task is not to fit into that dominant regime but to figure this unintelligibilty, this marginality, as a method of invention – a process, a 'rethinking' as Grosz calls it, that reorders the social, political and material culture that would strive to oppress it.

In Retallack's view this method of invention is 'feminine' and I would argue that it is also, by implication, spatial – a method that acts from a remote position, necessarily outside and tangential, that mobilises that position as an alternative place, not separate from, but active upon, dominant sites and discourses.

2   Joan Retallack, 'Re:Thinking:Literary:Feminism (three essays onto shaky grounds)', in Lynn Keller and Christanne Miller (eds), *Feminist Measures: Soundings in Poetry and Theory*, Ann Arbor, MI: University of Michigan Press, 1994.

5

She is rummaging in narratives, she's in a text on a page. She is looking for alternatives. Other writings buried in the material of these architexts. She works closely with the detail of this material, entering right into it. She's not afraid to get her hands dirty. It's a haptic and archeological reading that she is doing, and it gives her discourses to work with.

She has chosen not to site herself in full view but she is figuring her place as an opening, a 'motive' place, an open text. She chooses not to place herself in a position of fixity – she chooses (and that is most important) to confront the relationship to the other (the same), the viewer, the onlooker, the reader, the outside. And she confronts that relationship by not providing a stable body/place/sign to be seen/read against, by or as. But still a sign that is not 'empty', not EMPTY.[3]

And all the time she is also making decisions. Strategic, articulate, responsive decisions. She is negotiating narrative as space as well as time and moving as she encounters its discursive others. And she is aware of these narratives as fictions; delicately recomposed fantasies, fragments of referred language hinged together, scaffolds attempting to be facades.

This place, and who she is, are not a reflection of the closed narratives set out to construct her and the non-linearity of this open text, this place of writingbuilding, this 'diatecture', is where she finds herself. Singularity and closure mitigate against her, because she is figured by an opening, that is, by an action of opening into a space that is not merely reproductive but productive; a space of her that doesn't just nurture but also works.

3 See Rachel Blau DuPlessis, 'Whowe', in *The Pink Guitar: Writing as Feminist Practice*, New York: Routledge, 1990.

6

The term 'stray sod' comes from a story my grand-aunt used to tell me as a child. A man, on his way home from the pub one night takes short cut across the fields, and as he is walking he steps on a 'stray sod'. Suddenly he looses his orientation and finds himself spinning and dizzy. He collapses and when he opens his eyes he sees ahead of him the hill open up and a fairy wedding dancing out of it into the night. The story would vary from telling to telling, as stories do, but the essential thread was always the same; a short cut, stepping on a stray sod, dizziness and a vision.

Stray sods are unreasonable places; sites of land, of mind and of narrative. I imagine that all cultures must have a term that they use to describe these moments of 'drop out' where the usual everyday transmission of time and place is disturbed, where a sudden instance or discovery – a decision, an accident, a piece of information, pulls the ground from under us and we find that we have, as Seamus Heaney once described it, 'let down a shaft into real life'. Heaney is referring to his own poetry and his experience of the first time that he felt he had done more than just 'dabble in verses' and interestingly he talks about this creative moment as the 'first place where I felt that I had done more than just make an arrangement of words'.[4] The trap door that opened in his work, let him into the place of poetry and that too is a 'stray sod', a place and a loss of place, determined by something more than structure, vocabulary or intention, something strangely implacable – a dynamic perhaps, that makes its place in the inbetween of appearances and knowledge, that moves space – the ground of reason – into another register, another 'genre' almost, perhaps even (following Retallack) another gender.

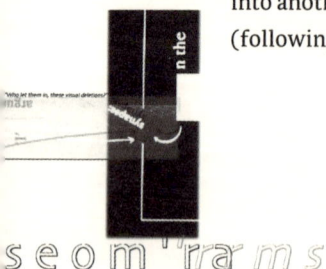

[4] Seamus Heaney, 'Preoccupations, Selected Prose 1968-1978', Faber & Faber, London 1984.

7

Many of the ideas in feminist writing about new constructions of the 'subject' propose a doubling up of the singular 'I' in order to produce a (feminine) subject that is, literally, hard to pin down. The doubleness of the maternal body, the 'sex that is not one', the stranger to ourselves, all identify a complex plural that substitutes for 'the', 'this', that replaces the definite article with a sign that is a 'shifter', a sign that is, in its indexical quality, its innate doubleness, inherently discursive, inherently spatial, inherently, I would say, 'feminine'.

'Shifter' is a term developed by Emile Benveniste (following Jacobson) to describe linguistic terms that depend upon the presence of their reference for meaning. They are terms such as 'this', 'that', 'here', 'there', 'I', 'you', 'now', 'then'. They only have meaning in an instance of discourse, being in themselves ciphers, absences through which language establishes location in time and space. The architecture of the shifters is the architecture of the index. They are deictic devices that show or point to the supposed real of which they are a copy or a substitute. Spaces that indicate other spaces – 'stray sods'. Their manner is to show, to point to, and to draw attention away from themselves. They are the double of a space itself, its shadow, its imprint, its now. And in this doubleness they provide a linguistic model for the contingent spatiality of the feminine.

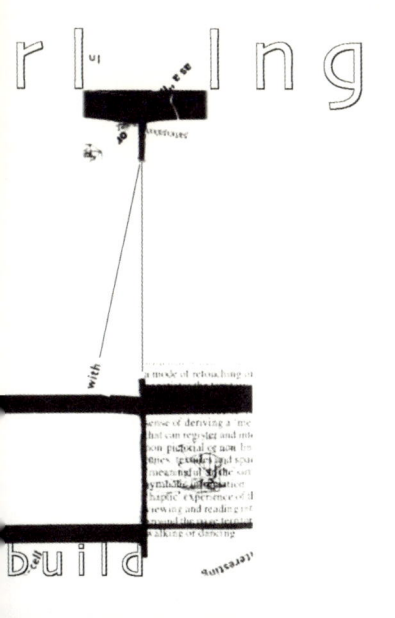

In her essay 'Developing the Negative', Peggy Phelan made a link between the doubleness of the feminine and the indexical doubleness of the photograph.[5] Both, she writes, 'are founded upon the terrain of the negative, both reproduce the visible'. And she says, 'To expose this reproduction is, perhaps, to ruin the security of the single subject endlessly seeking the perfect image s/he can (call her or his) own.' The dynamic of the index then prises open the notion of a singular subjectivity, a singular space and refigures the 'negative' feminine not as an absence, so much as an implacability – a stray sod that refuses to return the ideal image but offers instead a kind of paratopia – an alternative world, a space, inbetween things, 'in which things are undone, the space to the side and around, which is the space of subversion and fraying, the edges of identity's limits',[6] whose logic, always doubled, always this *and* that, is a 'depunctuated grammar' as Retallack would have it – reason let loose, reconstituting, rethinking and giving rise to new orders, new subjects, new spaces.

---

5 Peggy Phelan, 'Developing the Negative', in *Unmarked, Politics of Performance*, London: Routledge, 1993.

6 Elizabeth Grosz, *Architecture from the Outside, Essay on Virtual and Real Space*, Cambridge, MA: MIT Press, 2001.

8

Alterities, the conference, activated a place such as this. And over the intervening years, that place has reappeared in various ways, in different social, geographical and material forms. One such form occurred when one of the participants of *Alterities*, Katie Lloyd Thomas, and her friend, Teresa Hoskyns, both working in or around architectural practice in the UK, decided to organise an event to find out what work women architects were actually engaged in. They gradually invited a number of other women, myself included, many of whom had been participants at the *Alterities* conference, to join in this process of organisation. But as an all-female, cross-disciplinary and institutionally unaffiliated group (although many work in academia), 'the feminine', as other and outside and interruptiveness, gave rise to another kind of process. One which rather than organising an event – a finite moment – produced instead a subtle kind of interfering strategy – a method to intervene in social, institutional and perhaps other kinds of spaces. This group gave itself the name 'Taking Place' and became a feminine spatial method – a stray sod of sorts, slipped in under the feet of various forms of cultural, social and spatial legitimacy.

To date, 'Taking Place' has organised interventions/events at the schools of architecture at University of North London and Sheffield University. We have also published work in the UNL newsletter and the architecture journal *Scroope*, as well as presenting at 'Becoming Space: Ideology, Invention & Immanence in Human Surroundings', an international conference on 'Event Space', at the Living Art Museum, Rekjavik, Iceland.

The constitution of this group has changed in the few years that we have been active and it proceeds in a slow, response-led manner – taking on opportunities as and when they arise – and so it is a peculiarly contingent group and method, building and questioning through its disruption of other, already existing spatial contexts.

What is produced here is perhaps not so much an 'alternative' space, but an *altering space*, that which the feminine proposes as a disposition born out of a dispossession.

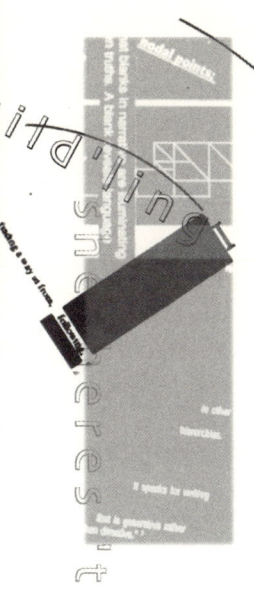

Chapter 8

Matrix; Chandra Mohanty
Paul Patton

Helen Stratford

M. E. Bailey; Gilles Deleuze and Felix Guattari

Irene Diamond and Lee Quinby

Philip Goodchild

Rosi Braidotti

MICRO-STRATEGIES OF RESISTANCE

Naomi Schor

Pauline Hanson and Geraldine Pratt

Sneja Gunew

Steve Pile
Patricia Waugh

Chapter 8 Helen Stratford

Fredric Jameson
Michel Foucault

Michael Waltze

Nancy Hartsock
Jos Boys

David Couzens-Hoy
Caroline Ramazanoglu and Janet Holland; Barbara M. Kennedy

Janet Wolff

Moira Gatens

GIES

Susan Bordo

Elisabeth Grosz

Jana Sawicki

Micro-strategies of resistance are particular confrontations with and resistances to the local impositions of dominating power. These incremental moves are not assembled from direct confrontations but rather operate as discrete traces within a plurality of resistances. Through a feminist critique, this chapter explores the implications that such an approach may hold for a feminist project, in terms of intersections of power relations and spatial practice.[1]

Micro-strategies: space and architecture

Resistance practices

The practice of all architects is to some extent an interpretation of the context in which they are located, whether this interpretation takes the form of tacit or explicit representation. This theoretical chapter has evolved from research which explores the work of various architects and groups who, in the early 1980s, questioned political and/or professional orthodoxies in architecture around them. The scope of that larger study stretches wide, geographically and politically, including a number of countries in political transition such as South Africa, Russia, Romania and East Germany. The groups studied include the Paper Architects, Utopica, Form-Trans-Inform and Matrix.[2]

The enterprises of the kind I have explored tend to be politically motivated since they respond to the social and/or cultural condition within which they are situated, even if the material itself cannot, by definition, be strictly political. As Frederic Jameson writes, 'No work of art or culture can set out to be political once and for all... for there can never be any guarantee it will be used in the way it

---

1 An earlier version of this chapter was given at the conference, 'Alterities: "feminine" practices of space', June 1999, l'Ecole Nationale Supérieure des Beaux Arts, Paris. All the components of this chapter are quotes, taken out of context and placed elsewhere. Their assemblage is a practice which invites re-readings, pointing towards a multiplication of meaning: a feminist approach which is not fixed and complete but rather a network of shifting, plural and non-linear relations.
2 See Helen Stratford, 'Micro-movements of Resistance: The Questioning of Orthodoxies by Young Architects in the East and West in the 1980s and Early '90s', unpublished dissertation for Diploma in Architecture, University of Cambridge, June 1998.

demands.'[3] In other words, even though a political reading can be made of the work, the work of art is 'in itself inert'.[4] In fact, in all the cases I have studied, it is less the architecture or the art 'form' which is deemed to be the ongoing location of protest than the actual *act* of creating them.

My search for these hidden practices was governed by a desire to study the response to an obliteration of opportunity, either in actual building processes or in wider forms of expression. Such a search can prove arduous. Those practices which do not fit the stereotypes are not understood by mainstream culture and the media where their message might be disseminated and can therefore be destined to obscurity. Further, the nature of resistance is not always necessarily overt or 'radical' but sometimes needs to be indirect: composed of subtle slippages and subversions.

For each case study, the socio-political context became the measure of the resistance tactic and the strength of the potential for resistance was aided by an understanding of the power relations which marked this context. In each case, the means through which certain power structures were stratified or set in place needed to be explored in order to gain an understanding of the relevance of their resistance strategy.

For the London-based feminist architectural collective Matrix, the domination of the building industry by men prompted a feeling of alienation and marginalisation and an awareness of the neglect of what they saw as women's needs in a 'man-made environment'.[5] Growing out of the late 1970s New Architecture Movement (NAM), of which many of the group had been members, Matrix began in 1980 and continued until its dissolution in 1996. In the context of the predominantly apolitical position of architectural practice and practices in Britain at that time, their book *Making Space* was one effort on the micro scale to raise awareness of this problem, asking how women – planners, architects, builders

---

3   Fredric Jameson, 'Is Space Political?' in Cynthia Davidson (ed.), *Anyplace*, Cambridge, MA: MIT Press, 1995, p.194.
4   Ibid., p.195.
5   Matrix were a group of feminist designers and a practice. Members and contributors included Jos Boys, Frances Bradshaw, Jane Darke, Julia Dwyer, Benedicte Foo, Sue Francis, Barbara Mcfarlane, Marion Roberts and Anne Thorne. See also Julia Dwyer and Anne Thorne's chapter in this book.

and clients – might take part in the making of space.[6] While providing a key commentary on issues around the expression and production of inequality through language, representation and practice, this book documented a multitude of interventions conducted at the level of everyday activities including projects for women's communities, children's institutions and immigrant groups.

Through workshops, exhibitions and conferences as well as built projects, Matrix's practice subverted and challenged the traditional relationship between architect and client. In place of the independent, elitist and gendered role of the Architect, the practice focused on full participation in the design process from women who would eventually use the end product. Various methods were used, including sessions about a building's programme and how to make and read architectural drawings, to create situations through which women might attain a voice in the construction of the built environment.

The range and character of these practices and those of the other case studies, located at the level of everyday life and frequently discrete and hidden, informed the theories of resistance that linked all the separate case studies together. This 'theoretical' thread drew upon the work of Michel Foucault to question conventional notions of power and resistance.

Theories of power/space

It is exactly the notion of micro-strategies composed of slippages and subversions, assembled from the materials and practices of everyday life, that so strongly resonates with the views of Michel Foucault.[7] In *Power/Knowledge*, Foucault comments:

> Power is employed and exercised through a netlike organisation. And not only do individuals circulate between its threads; they are always in the position of simultaneously undergoing and

---

6 Matrix, *Making Space: Women and the Man-Made Environment*, London: Pluto Press Ltd, 1984.

7 In *Geographies of Resistance*, Steve Pile describes how resistance practices can be 'tiny micro-movements of resistance'. 'Micro-Movements of Resistance' was the former title of this chapter and the title of the dissertation which provided the practical case studies from which this chapter has been theorised; see note 2 above. Steve Pile and Michael Keith (eds), *Geographies of Resistance*, London: Routledge, 1997.

exercising this power. They are not only its inert or consenting target, they are always the elements of its articulation.[8]

Hence, for Foucault, power is never wholly expressed on a global scale, only at local innumerable points as 'micro-powers' in an 'endless network of power relations'.[9] Foucault explains that 'the overthrow of these micro-powers does not obey the law of all or nothing... There is a plurality of resistances, each of them a special case.'[10] In this way, as David Couzens-Hoy astutely observes, 'Change does not occur by transforming the whole at once but only by resisting injustices at the particular points where they manifest themselves.'[11]

Through the panoptic mechanism, Foucault famously extends his notion of power to the power/space relationship. However, this relationship cannot be specifically restricted to *architectural* space. By conjoining bodies and space, Foucault's panopticism is a 'generalised function': 'a way of making power relations function in a function and of making a function function through these power relations'.[12] Power becomes for Foucault a realm that is not deterministically connected to space yet inextricably depends upon it.

Architectural space cannot dictate use – there will always be a resistance of use at the level of the everyday: 'the architect has no power', says Foucault.[13] Nevertheless, in their relations and connections, spatial layouts, representations and practices can facilitate and support, to some extent, a politics of use within a specific socio-political context. Consequently, as Foucault comments, spatial layout 'can and does produce positive effects when the liberating intentions of the architect coincide with the real practice of people in the exercise of freedom'.[14] Ultimately, space is *never* free from its contextual

8 Michel Foucault, *Power/Knowledge: Selected Interviews and Other Writings, 1972-1977*, Brighton: The Harvester Press, 1986 p.98.
9 Michael Waltzer, 'The Politics of Michel Foucault', in David Couzens Hoy (ed.), *Foucault: A Critical Reader*, Oxford: Basil Blackwell, 1986, p.55.
10 M. Foucault, cited in Waltzer, 'The Politics of Michel Foucault', p.55.
11 David Couzens-Hoy, 'Power, Repression, Progress', in *Foucault: A Critical Reader*, p.143.
12 M. Foucault, *Discipline and Punish: The Birth of the Prison*, New York: Vintage/Random House, 1979, pp. 206-7.
13 M. Foucault, 'An Ethics of Pleasure', in Silvère Lotringer (ed.), *Foucault Live (Interviews, 1966-84)*, New York: Semiotext(e), 1989, p.267.
14 M. Foucault, 'Space, Knowledge and Power', in Paul Ranibow (ed.), *The Foucault Reader*, London: Penguin, 1991, p.246.

politics, *never* neutral. Most people live within a specific contextual field of space and spatial relations. In 'Foucauldian Feminism', M.E. Bailey describes how 'ideas of the "feminine" are the result of the interplay of previous historical understandings of femininity and the bodies these have produced', in a way comparable to that in which 'ideas and materiality are intertwined in a spiral of mutually informed contingency', for Foucault.[15] Therefore, one can presumably posit a resistance that prevents the coercion of a detrimental politics of use as a blockage, a subversion that, in its liminal position, relies on incremental moves rather than an overt formal strategy.

Micro strategies: a feminist project?

Difference

The idea of a resistance, which represents the interests of a group that makes up over half the natural population of a country, can bring with it obvious erroneous assumptions and contradictions. There is a risk of arguing that the experiences concerning a history of gender oppression through sex/gender hierarchies are shared by all women, transcending divisions created by race, age, geographical location, social class, parental status, health, religion, ethnicity and sexual orientation. Indeed, the impulse to blend into an undifferentiated 'sisterhood' has been interpreted as 'a feminist version of imperialism'.[16] Moreover, as some feminist theorists have observed, 'the concept of a "woman's identity" functions in terms of both affirmation and negation', and if the essentialism of this concept is embraced, it risks reproducing the very patriarchal construction of gender which is being contested.[17] So, is Foucault's model of power/spatial relations useful for feminist praxis?

---

15 M. E. Bailey, 'Foucauldian Feminism', in Caroline Ramazanoglu (ed.), *Up Against Foucault: Explorations of Some Tensions between Foucault and Feminism*, London and New York: Routledge, 1993, p.104. See also Sandra Lee Bartky, 'Foucault, Femininity and the Modernization of Patriarchal Power', and Susan Bordo, 'Anorexia Nervosa: Psychopathology as the Crystallization of Culture', in Irene Diamond and Lee Quinby (eds), *Feminism and Foucault: Reflections on Resistance*, Boston: Northeastern University Press, 1988.

16 Sneja Gunew, 'Authenticity and the Writing Cure: Reading Some Migrant Women's Writing', in Susan Sheridan (ed.), *Grafts: Feminist Cultural Criticism*, London: Verso, 1998, p.115.

17 Patricia Waugh, 'Modernism, Postmodernism, Gender', in *Practising Postmodernism/Reading Postmodernism*, London: Edward Arnold, 1992, p.119.

Diffuse

Foucault's model of power relations has been severely criticised by feminists for a levelling out of gender relations. Nancy Hartstock and Nancy Fraser have argued that, through dismissing gender configurations of subordination to power grounded in patriarchal relations, Foucault's conception of power in itself is gendered. Power is everywhere and so ultimately nowhere.[18] Also, Caroline Ramazanoglu and Janet Holland take Fraser's line in arguing that, for Foucault, power becomes 'such a "catch-all' category that it leaves him unable to distinguish between the exercise of different kinds of power, and so unable to judge between the morality of different kinds of constraints'.[19]

However, as other feminists such as Jana Sawicki and Moira Gatens have observed, it is *exactly* this diffuse model that has significant implications for a feminist project in its recognition of the more insidious and non-repressive, but generative, forms of power.[20] As Bailey argues, the collapse of the 'primordial monolith of "patriarchy" into the fragmented and diffuse, but undeniably interlocking, specific structures of masculinist power' criticises implicit or explicit essentialism in feminist discourse. It 'frees feminists to pursue specific local struggles without justifying these with reference to an entirely male system of power and consequent oppositional female powerlessness'.[21]

Multiplicity

Foucault's plural and partial identities point towards what Sawicki describes as a 'shifting politics of difference'.[22] Here, according to Chandra Mohanty, there is no 'universal category of woman', but shifting spatial and temporal commonalities are produced at the intersections of various networks of power

---

18 See Nancy Hartsock, 'Foucault on Power: A Theory for Women?' in Linda Nicholson (ed.), *Feminism/Postmodernism*, London: Routledge, 1989, and Nancy Fraser, 'Foucault on Modern Power', in *Unruly Practices: Power, Discourse and Gender in Contemporary Social Theory*, Cambridge: Polity Press, 1989.
19 Quote of Fraser's 'Foucault on Modern Power', in Caroline Ramazanoglu and Janet Holland, 'Women's Sexuality and Men's Appropriation of Desire', in *Up Against Foucault*, p.256.
20 Cf. Moira Gatens, 'Power, Bodies, Difference', in *Imaginary Bodies: Ethics, Power, and Corporeality*, London: Routledge, 1996, p.66.
21 M. E. Bailey, 'Foucauldian Feminism', p.119.
22 See Jana Sawicki, *Disciplining Foucault: Feminism, Power and the Body*, New York and London: Routledge, 1991.

relations.[23] Also, as Couzens-Hoy observes, Foucault resists any thinking rooted in totalising viewpoints and he maintains that 'there is no such thing as power as a whole, and no standpoint from which the totality could be viewed or evaluated'.[24]

Ultimately, if Foucault himself glosses over gender configurations of power, his ethics, grounded in a resistance to whatever configuration *totalitarian* power might take, can prove relevant for feminists in contemporary society where the two regimes of dominating and generative power coexist and often intertwine.[25] Indeed, as Couzens-Hoy concludes, if for Foucault there is no social existence *without* power relations, this does not entail that particular, oppressive power relations are necessary. Furthermore, his rejection of the notion of universal progress does not necessarily 'abandon the hope for emancipation, if by that one means the resistance at particular points to local exercises of power'.[26] Here then, resistance is not restrained to immutable boundaries, it aims to search for pertinent and progressive ways of considering the fluidity of boundaries among people which make it possible for difference to be embraced.

Scale and intimacy

Foucault's contention that the large-scale organisation of power is connected to the most minute and local practices witnessed at the detail and intimate levels of everyday life, not least at the site of the body, align with feminist understandings of the 'personal as political'.[27] It holds particular relevance to those people who traditionally have been confined to the most local and personal realms of life. Here, at the extremities of the network, its effects are felt most severely. It is in the discourses of these marginalised groups, suppressed by rationalising discourses, which lay claim to reason and ultimately truth, that Foucault sees sites of resistance.

---

23 Chandra Mohanty comments how the intersections of 'various systematic networks of class, race, (hetero) sexuality, and nation... position us as "women"'. See 'Introduction', in Chandra Talpade Mohanty, Ann Russo and Lourdes Torres (eds), *Third World Women and the Politics of Feminism*, Bloomington, IN: Indiana University Press, 1991, p.13.

24 D. Couzens-Hoy, 'Power, Repression, Progress', p.142.
25 See Irene Diamond and Lee Quinby, *Feminism and Foucault*, p.xiv.
26 D. Couzens-Hoy, 'Power, Repression, Progress', p.145.
27 See J. Sawicki, 'Identity Politics and Sexual Freedom', in Diamond and Quinby (eds), *Feminism and Foucault*, pp.186-7.

From a personal position of a minority within the architectural profession, Matrix understood more than other architects the effectiveness of micro-powers as pervasive agents of control expressed at the level of everyday life.[28] In *Making Space*, Jos Boys describes how 'the emphasis on individual mobility through private car ownership since the second world war' and the zoning aspect of post-1945 town planning were reflected in the way 'man-made surroundings lacked consideration for the less mobile'.[29] At a time when multiple car ownership was uncommon and the socially approved role for women was the major child carer, this resulted in isolation for women and a 'lack of mobility' at the level of daily activities.[30]

In 'Power, Bodies, Difference', Gatens describes how Foucault's 'micro-politics' are particularly pertinent to 'an investigation of the ways in which power and domination operate in relation to sexual difference'. She goes on to describe how 'social practices construct certain sorts of body with particular kinds of power and capacity'.[31] Likewise, Susan Bordo has explored how societal practices mark bodies through normalising processes, dominant cultural forms such as 'sexism, racism and ageism', which carry with them notions of 'acceptable' and 'abject' bodies.[32] Gatens describes how it is this marking which, in turn, 'creates specific conditions in which [these bodies] live and recreate themselves'.[33] In a later paper, she draws upon the work of Guattari, Deleuze and Spinoza to expand upon the processes involved in the construction of sexual difference. She describes how the historical relation between women and men may be read as 'clusters of specific affects and powers of bodies which are organized around an exclusive binary form (male/female), through various complex assemblages', social, environmental, cultural, legal and linguistic.[34]

---

28 Matrix drew attention to the fact that in 1984, 95% of architects were men. See *Making Space*, p.2.
29 Jos Boys, 'Women and Public Space', in Matrix, *Making Space*, p.41.
30 Ibid.
31 M. Gatens, *Imaginary Bodies*, p.67.
32 Susan Bordo comments that such dominant cultural forms as 'Sexism, Racism and "ageism", while they do not determine human value and choices, while they do not deprive us of "agency", remain strongly normalising within our culture'. Susan Bordo, 'Feminism, Foucault and the Politics of the Body', in *Up Against Foucault*, p.199.
33 M. Gatens, *Imaginary Bodies*, p.71.
34 M. Gatens, 'Through a Spinozist Lens: Ethology, Difference, Power', in Paul Patton (ed.), *Deleuze: A Critical Reader*, Oxford: Blackwell, 1996, p.178.

Molecular

The notion of subjectivity as a social phenomenon, defined on the periphery rather than at the centre, as exteriority rather than interiority, is a key concept in the work of Guattari and his collaborations with Deleuze. Like Foucault, Guattari is concerned with 'the microscopic means of disciplinarisation of thought and affect and the militarisation of human relations'.[35] In the first chapter of *Chaosmosis*, Guattari argues for a complexity of subjectivity. He opens the production of subjectivity up to 'machines of subjectivation',[36] which are not delimited to 'internal faculties of the soul, interpersonal relations and intra-familial complexes', but are instead found in 'non-human machines such as social, cultural, environmental or technological assemblages'.[37] For Guattari, these include 'the large-scale social machines of language and the mass media'.[38]

By attacking what he terms the countless varieties of 'micro-fascism' which are accepted through uncritical participation in hegemonic structures of power, Guattari targets the myth of the inevitability of hierarchy and authority. The possibility of 'other types of subjective arrangements', not based on relations of power, is dependent on 'an analysis of unconscious relations, of relations of micro-power, of micro-fascism'. It is focused on the struggle against the formation of the subject in 'constant power relations, in relations of alienation and in repression against bodies, thoughts and ways of speaking'. Guattari terms this reviewing of subjectivation on the micro-level 'micro-politics'.[39]

Micro-politics

While Foucault's 'micro-politics' might run the risk of provoking passivity in the face of dominating and omnipresent

---

[35] Felix Guattari, *L'Inconscient machinique: essays de schizo-analyse*, Fontenay-sous-Bois: Encres/Recherches, 1979, p.200, quoted in Gary Genosko (ed.), *The Guattari Reader*, Oxford: Blackwell, 1996, p.23.
[36] F. Guattari, *Chaosmosis: An Ethico-Aesthetic Paradigm*, trans. Paul Bains and Julian Pefanis, Bloomington, IN: Indianapolis University Press, 1995, p.9.
[37] See P. Goodchild, *G. Deleuze and F. Guattari, An Introduction to the Politics of Desire*, London: Sage, 1996, p.15.
[38] F. Guattari, *Chaosmosis*, p.9.
[39] Guattari comments, 'The micro-political struggle of desire cannot be fought out on any front without determining first the machinic-semiotic composition of the power formations.' See 'The Micro Politics of Fascism', in *Molecular Revolution*, trans. Rosemary Sheed, London: Penguin, 1984, p.224.
[40] P. Goodchild, *Deleuze and Guattari*, p.151.

power structures, to which resistance is only possible at particular points, Guattari's project specifically looks for sites of weakness in the structures themselves. For him, it is a question of expanding the assemblages that enter into the production of subjectivity in order to 'produce new kinds of subjectivity and new states of consciousness', 'lines of flight' which offer possibilities for change and movement.[40]

In a feminist context, Barbara M. Kennedy describes how 'micro-politics is an engagement with thinking about how struggles occur not only in groups, but through and even beyond multiplicities of identities, subjectivities and corporealities, internal to or functional through and across individual subjects and within subjects'.[41] In fact, this understanding of 'micro-politics' finds allegiance with many political positions including feminism. In 'A Thousand Tiny Sexes', Elizabeth Grosz describes how

[the] affirmation of localised, concrete, nonrepresentative struggles, struggles without leaders, without hierarchical organisations, without a clear-cut program or blueprint for social change, without definite goals and ends, confirms, and indeed, borrows from already existing forms of feminist political struggle, even if it rarely acknowledges this connection.[42]

As Kennedy describes, 'micro-politics is concerned with the micrology of lived experiences, across and between the spaces of any fixed, sentient or even fluid gendered subjectivity'.[43] This idea of the human being as 'multiply constituted' and 'traversed by diverse social formations' is taken further in terms of a specifically feminist agenda by the 'nomadic subject' of Rosi Braidotti. Here, the materialisation of bodies is not simply a passive inscription. In fact, the conception of human being as a component of a dynamic inter-related aggregate makes possible the potential to express being otherwise.

41 Barbara M. Kennedy, *Deleuze and Cinema: The Aesthetics of Sensation*, Edinburgh: Edinburgh University Press Ltd, 2000, p.21
42 Elisabeth Grosz, 'A Thousand Tiny Sexes', in Constantin V. Boundas and Dorothea Olkowski (eds), *Deleuze and the Theater of Philosophy*, London: Routledge, 1994, p.193.
43 B. Kennedy, *Deleuze and Cinema*, p.21.

Nomadic

If Foucault and Guattari's work is useful to feminism, Braidotti's concept of 'nomadism' seems to encompass these ideas from a specifically feminist perspective. For Deleuze and Guattari, the idea of nomadism expresses an alternative mode of thinking that operates beyond the conceptual structures produced by and productive of the established order or the State.[44] Braidotti's reading of nomadism politicises that of Deleuze and Guattari. For her, it offers a different way of thinking through the female feminist subject in terms of three simultaneous levels: (1) 'differences between men and women'; (2) 'differences among women'; and (3) 'differences within each woman'.[45] By focusing on the 'becoming of the subject' split over 'multiple axes of differentiation', such thinking resists 'assimilation or homologation into dominant ways of representing the self'.[46] It locates political agency in 'nomadic engagements' which shift between all levels and work through the 'stock of cumulated static images, concepts and representations of women, of female identity'.[47]

Feminist geographers such as Geraldine Pratt and Pauline Hanson have observed how 'gendered, racialised and classed identities are fluid'. They are constructed and sustained by geographies of place, space and time, but are produced in 'different ways in different places'.[48] Nomadism recognises the fluidity of these discontinuous locations. It implies both the necessary complication of the term *woman* and a constant reassessment of one's location in power structures. A continual re-mapping must take place on multiple levels associated with different locations in time and space and different economic and social practices, so as to posit an effective resistance against, and to move beyond, hegemonic constructions of gender which attempt to fix ideas about people in time and space.

Nomadism resists fixity and closure, but the idea of movement which is inherently liberatory is problematic.

44 See Gilles Deleuze and Felix Guattari, *A Thousand Plateaus*, London: Athlone Press, 1988.
45 Rosi Braidotti, *Nomadic Subjects: Embodiment and Sexual Difference in Contemporary Feminist Theory*, New York: Columbia University Press, 1994, p.158.
46 Ibid., p.25.
47 Ibid., p.169.
48 Pauline Hanson and Geraldine Pratt, 'Geography and the Construction of Difference', *Gender, Place and Culture* 1.1 (1994): 9-12.

Women traditionally are defined within specific *fixed* places, whether literal or ideological, while the concept of fluidity in respect to femininity carries the notion of darkness, mysteriousness and decadence. Nonetheless, the acknowledgement of the identity of the 'nomadic subject' as fluid might serve to point towards a deconstruction of those static dualisms on which such androcentric notions of decadent female fluidity are based. As Janet Wolff has observed, the *reappropriation* of metaphors here becomes fundamental. The *specific* utilisation of such concepts as 'movement' and 'fluidity', by whom and within what particular context, determines whether they promote resistance to hegemonic functions and formations or whether they become their own academic reification and tools of hegemony themselves.[49]

Micro-strategies: examination of the detail

The traditional way of implementing architecture begins not at the minutiae but at the world's eye view. The architect stands above the site plan, reduced to convenient drawing board scale, and bequeaths to the world a heroic edifice. The topography of the micro-site involves a shift in scale from the universal to the specific which, like Foucault's genealogies, may recover a different perspective. To begin from the detail as a way of thinking through the 'micro', to begin from the realm of the intricacies of everyday life, might posit a way of offering a different way of thinking through architecture. Such an approach may focus on steps of inclusion which recognise a rich texture of difference, proposing points of connection rather than a totalising and exclusionary move.

In initiating localised counter-responses to everyday processes of stratification, micro-strategies operate at a molecular level. The introduction of the concept of micro-strategies to architecture aims to lead to a productive and critical reformation of the conventional ways of perceiving architecture. Here the 'molecular' does not oppose the intervention on a large scale but rather refers to the rejection

---

[49] Janet Wolff, 'On the Road Again', *Cultural Studies* 7 (1992), pp.224-39.

of any general or universalising emphasis, indicating 'not a difference in scale but a difference in kind'.[50]

Ultimately, this reviewing of space, this movement composed of micro-resistances is neither located in the formal expression of the end building, nor constrained to the detail. Instead, it constantly shifts between all fragments of the building process to unveil local exercises of hegemonic power. In this way, these micro-points become the starting point for the means of expression for those whose space and discourses are circumscribed by the fragmented and diffuse, yet interlocking structures of masculinist power.[51]

Whether in the form of the 'ornamental' as effeminate and decadent or in the everyday as the domestic sphere of social life, as Naomi Schor describes in *Reading in Detail*, the detail has long been aligned with the feminine. However, it is *how* the micro is viewed and through *what* lens which is key. Micro here then is perhaps less a specific scale of looking than a *field of inquiry*. A dispersed framework through which to re-view space and spatial relations. Larger general terrains may be (re)viewed through the micro. As Schor comments, 'To retell the story from the perspective of the detail is inevitably to tell another story.'[52]

---

50 Paul Patton, *Deleuze and the Political*, London: Routledge, 2000, p.43.
51 Matrix's project, Jagonari Educational Resource Centre in Whitechapel, in which they worked with women clients/users on decorative components for the outside of the building, as ways of representing cultural and gender difference through 'detail', might be an example in this sense.
52 Naomi Schor, *Reading in Detail: Aesthetics and the Feminine*, New York: Methuen, 1987, p.4.

Acknowledgements
I am grateful to Dr Jane Rendell for advice and direction on this chapter. Likewise, I am grateful to Doina Petrescu for organising the 'Alterities' conference, creating a place to challenge the current boundaries of the field of architecture from a feminine and interdisciplinary perspective.

Chapter 9

Anne Querrien

ALTERING EVENTS
IN ARCHITECTURE

Donna Haraway

Jean Viard

Michel Foucault

Chapter 9   Anne Querrien

Jean Patrick Fortin

Félix Guattari

Jürgen Habermas

Gilles Deleuze
Buster Simpson

The sight of 'the other' in modern architecture

Working as a sociologist and an editor of urban research and trying to help my readers to visualise the spaces that we speak of in our publications, I am always surprised to receive from architects images showing buildings without any human being around. I have been told that modern architects do not want to be explicit about the real scale of what they build, and yet landscape painters have done exactly the contrary, by always drawing tiny figures, to allow the appearance in the background of the town or the monument in all its magnificence. But that was drawing or painting and not photography; and as such, contained no objective mechanism in charge of giving the 'right' dimensions between bodies and buildings. Within those representational techniques, dreams were still allowed.

And even stranger to my eyes is when sociologists or urban planners take snapshots of the sites they research and let you think that a new kind of desertification is produced by the power constraints on the territory. A new building, even if it is social housing, comes from nowhere in their pictures, as a completely new occupation of the land, without any trace of local users. The pictures in which users are represented are mostly those that show them linked to forms of 'resistance': occupants of contested spaces gathered in demonstrations or exemplifying lifestyles that are going to disappear. Most of the time women 'take the floor' in these pictures, as in movies 'the speech', to emphasise all the habits which it is not possible to keep in more modern settings.

Invisibility of women and the everyday life

Washing and drying out laundry have been eliminated step by step from the open spaces of residential areas because they were considered 'unsightly'. Regulations in the collective housing estates prohibit hanging out laundry to dry at the front windows or in the front yards, an interdiction that has also been extended along roads and pedestrian routes. Drying clothes outside has become impossible in the inner city, remaining a practice accepted only in the suburbs.

In Le Corbusier's buildings and in other modern architectures, all walls are front walls.

The courtyard of the working-class estates of the nineteenth century has disappeared from the modern estates, and with it, the small public place for women's speech. The street corner laundrette could eventually become a place for talk but it is too far from the kitchen, which gives direct access to the yard and allows the movement of going in and retiring from the public sight, following the waves of washing and drying. The washing pool in the village and the common washing basin in the courtyard were places for women's socialisation, gossip and solidarity.

While raising their eyes to the clotheslines hung across the streets in the South, the tourists are delighted at the sight of all these prohibited habits that continue there, where women's domestic work is still visible in the public space, like all the other little street activities connected to everyday life. Here the ordinary houses have not yet become pieces of architecture, and the alterations of ordinary life are still welcome, as a kind of narrative of what is going on. In the North, the architectural regulations have been written to follow the disciplinary work of the factory. Here the neighbourhoods, ruled by factory leaders and municipalities under their control, were strictly ordered according to the work hierarchy and could not be altered by the marks of persons 'who do not work' (i.e. women, children, animals and others). All activities accompanying everyday work have become private, forced to remain invisible and contained within the private space. In the industrial town, public space has been restricted to functional 'ways': work-way, school-way, store-way, none of which cross each other. But in the post-industrial town, massive unemployment brings with it a kind of urban revolution, claiming for a new interpretation of public space, claiming new landscapes and spaces for hanging out: spaces that were previously assigned only to the 'leisure' classes and privileged populations.

Making life invisible and conceiving of building as a still landscape seems one of the aims of modern architecture, understood as an architecture 'affordable to the working class', but also an architecture meant to push people out to work. Economic incentives are most effective for such an objective:

'if you have to pay more and more for your home, then you try to earn more and more'. The home features no longer matter so much but rather the relationships to the transport system, the shopping areas, the schools for children, and the security within any kind of supply, the 'job' being the first one. Architectural quality does not seem a priority; the new urban developers, in search of maximum financial profits, tend to offer for sale what has proven to work in the past. Architecture has become a pastiche of itself, a barrier to innovation, not only within its own domain, but also in all social fields.

Does social mix alter the 'generic city'?

This reductive and homogeneous understanding of architecture that leads to what Koolhaas has called 'generic cities', meaning to 'envelop' everything and keep public space out of the grip of others, cannot stand the social complexity of the global city and its mixed populations coming from different cultures, engaged in different economies with different temporalities. Except for the privately managed places of tourism, public space cannot afford the smoothing imposed by the management of the contemporary city, which wants to let us think that everything and everybody is working, while work never represents more than 15 per cent of a lifetime.[1]

So everywhere now, architecture, and especially, housing, need regeneration. New urban projects try to realise a smooth city in which dirty jobs are no longer visible and the equality between residents is understood in terms of their common ability to afford private ownership. Most of these projects are developed on derelict estates in the post-industrial areas inhabited mainly by the former working class, to indicate that what is needed today is not the extension of the city, but an efficient way of rebuilding and redistributing it from inside. All citizens will eventually enter the process and have to move, adapting their old habits and creating new ones in new settings and thus obeying the new architectural order. It begins with nearly nothing: an electronic device and a number

---

1   Jean Viard, *Le sacre du temps libre: la société des trente-cinq heures*, La Tour d'Aygues: Editions de l'Aube, 2002.

at the door, and the plot will no longer be crossed. You are in a private property, and like the others who share it with you, you must believe in its 'generic' unity and its gated community...

Making invisible the differences between people, trying to integrate them into normative common uses, could seem like a democratic and egalitarian project.[2] Yet this normalisation duplicates the former differences, transforms them into more sensitive differences and generates new conflicts. This impossible assimilation turns into a multiplicity of small differences and creates new social demands. The dominant power supported by the new urban developers cannot understand that what is perceived as 'resistance' to urban renewal (i.e. rejection of regeneration schemes, urban actions, etc.) is the consequence of its own difficulties in trying to create a 'democratic' urban space. Those who have moved inside this smooth and clean new world cannot afford the new consumption they are forced into, nor are they able to find the open life they have dreamt of.

An interesting case is that of the Alma Gare neighbourhood in Roubaix, a small post-industrial city in northern France. This regeneration project conducted in the 1970s which has became a well-known example of community participation in France, proposed a shift from the traditional working-class courtyards to a new social housing scheme.[3] Social workers and neighbourhood associations thought it would be better to improve the old single-storey houses with technical innovations than to go for high-rise buildings deprived of collective life; but the result was almost the same. The new settlement was designed with the working-class residents and tried to repeat the space of the former estate but without collective washing facilities and other urban amenities that were considered obsolete. In order to afford the new modern comforts, the former working-class dwellers were forced to find other jobs in order to pay for the maintenance of their renovated flats. To enhance the family income, women where pushed

2 See Jean Patrick Fortin, *Grands ensembles, l'espace et ses raisons*, Paris: Plan Urbanisme Construction Architecture, 1999.

3 See also A. Querrien, 'How Inhabitants Can Become Collective Developers: France 1968-2000', in P. Blundell Jones, D. Petrescu and J. Till (eds), *Architecture and Participation*, London: Spon Press, 2005, pp 105-15.

into working as well and were no longer available to spend time together and share the collective amenities and the courtyards which, little by little, became the realm of dealers and youth threatening the elderly inhabitants.

What for a long time was considered the main reference for participative architecture in France has now ended in demolition, similar to many other social housing estates in Europe. This participative process in the Alma Gare project did not integrate the new type of residents and their lifestyles, but insisted on organising the renewal around the mythic figure of the working-class dweller. This quite homogeneous neighbourhood formed by residents formerly working in the iron and textile industries, in the past took advantage of very cheap old housing estates with plenty of common space and strong habits of solidarity. As soon as they were settled in their new comfortable flats, they began to dissociate from each other, and embrace a new lifestyle. Apart from the new comfort and the need for money to pay for it, other changes were mainly provoked by the newcomers in the neighbourhood, who were mostly immigrants following an upwardly mobile social route. As in many social housing estates in France, this change within the social composition of the neighbourhood was not easily accepted by the old inhabitants and generated conflicts.[4] Consequently, the architecture that was ideally intended to host a mixed population has now become the symbol of a remarkable failure.

In addition to the transformed social housing estates in the cities, the small family houses of the suburbs share the same spatial characteristics: individual comfort, enclosure of everyday life within private space resulting in the complete separation between the private and the public and the desertification of the public space in which the presence of the other is always feared. This is not a real alternative for anyone, and certainly not an alternative for the poorest who, if they cannot afford to keep living in their existing social housing estates, have to move to a new kind of place, which is better controlled socially. What is then to be done?

4 Nevertheless, the working-class composition in France is today mostly immigrant, as the old working classes have pushed their children ahead, in the service industry or in technically qualified jobs.

The power to reproduce difference

Can we dream of alternative spaces, which could be built in these derelict places, either public or private? This is only possible if the owners are willing to try a new experience, perhaps only temporarily, which could add value to the place and lead to the desire for another use. Such a space will not try to make difference invisible, but, instead, to register it by signs that are collectively decided upon in the experiment with all participants, whatever their status; that is to say the participants 'here' and 'now', who are present at the moment when the decision is taken. It would be an experiment in *building a public space together*, both at the political level, as a 'deliberative space', to pick up Habermas' formula,[5] and at the physical level, as a self-managed space.[6] This space would not reduce the needs of its users – which could include companies, services, or ordinary people – to a common norm of functioning and appearance, as was the case with modern architecture. The integration of local differences into a landscape of differences, the representation of the local and its relation to the global would become part of the architect's new social and spatial practice. Difference would appear as a power to reproduce difference, to welcome the new as different, as a new hybridising power.[7] Architecture would become an experiment in building together using the knowledge and skills of the architects. People who participate in such experiments will be able to take ideas for their homes, for the public spaces in their cities and neighbourhoods, and use their own means and their own words, with perhaps some help from architects. Such participation of architects in ordinary life would of course need a new economy of the architectural profession that could evolve into a collective service, even a public service in some cases.

Such a different agency between architects, specialists and users would not probably result in vastly new products in the beginning, as everybody would still be too impregnated by the imposing norms of currently practised architecture.

5   See Jürgen Habermas, *L'espace public*, Paris: Payot, 1978.
6   See *Territoires*, 431, profile issue, 'Luttes urbaines et autogestion', October 2002.
7   See Gilles Deleuze, *Différence et répétition*, Paris: Presses Universitaires de France, 1968.

Yet the important feature of such an experimental space is that it could be a real space practised for its own sake, and not for a predetermined function. It is important to experience explicitly the fact that space is more than function. The error of modern architecture was to believe that 'function' comes first and has to be conserved by any means possible. The collective laundry in the village is a place to gather, to sing, to gossip; it is a place distant from the house, a place to walk to in order to meet others. On the lines on which I hang out my laundry to dry, I also unfold my taste, my wealth, I display myself: it is a place to say publicly who I am. It is something I cannot do alone in my kitchen. In the contemporary city it is more and more difficult to have a real presence in public space as an individual, to have a public place of your own.

The end of disciplinary space

Donna Haraway has underlined the situatedness of knowledge, which is always rooted in life experience, which cannot be the same for women and men, for former colonised and former coloniser, and even for women of different social milieus.[8] The modern architectural knowledge was formed in the Renaissance, together with the drawing techniques that allowed architects to master the work on space and to control the building skills of workers. Classical and baroque styles gave an aesthetic motivation build for others, for other social classes and people from other origins, came later. The disciplinary power of space expressed in the architecture of prisons, schools, military barracks, factories and hospitals at the end of the eighteenth century was not conceptualised before the end of the nineteenth century. Foucault has shown that the disciplinary power of the Panopticon is found in the prisoners' conviction that they could be seen at any moment by the guard.[9] The social relation is then shaped by the potential menace of the glance, not by its actuality, which may never happen.

---

8   See Donna Haraway, 'Situated Knowledge: The Science Question in Feminism as a Site of Discourse in the Privilege of Partial Perspective', *Feminist Studies*, 14(3) (1998).

9   Michel Foucault, *Surveiller et punir*, Paris: Gallimard, 1975.

The idea that the spatial envelope is a disciplinary device in itself and can work to serve a functional way of living emerged with the early capitalist industrialisation and urbanisation and continued within modern architecture. The house was subordinated to the working place and located close to it, being linked to and organised by the working conditions. In industrial cities, the size of house varied according to rank in the factory, as if human needs were sized according to the same hierarchy.

Work is no longer today the centre of life; the diversity of residents and their preferences, the taste for experiment and debate are the condition for the creation of new lifestyles and living experiences in the former industrialised neighbourhoods. The 'other' is no longer a superseding 'voice' or 'eye', but a multiplicity of whispers and proposals coming horizontally from all directions and waiting for hybridisation and creation. The scale could be small and the conquest of visibility a first stage in inventing a new lifestyle.

Allowing women and men to dry their laundry at their windows, once more could be a sign of 'poetic utility', to use Buster Simpson's term. His urban installation, 'Shared Clotheslines: Banners of Human Reoccupation', realised in 1978 on Solar Day, in downtown Seattle, connected four floors and their inhabitants with their neighbours across the alley through clotheslines.[10] This urban amenity (realised at about the same time as Alma Gare project), which proclaimed sustainability as well as social issues, 'served as an alternative to the electric dryers as well as a response to the proliferations of the decorative "identity package" banners marketing a lifestyle rather than providing an authentic urban experience'.[11]

Allowing such an urban experience could also prove that a lot of windows are now without women or men acting as women. How much of our built environment is still left in our hands? Drying laundry at the windows of all kinds of houses and buildings could be, as in Simpson's case, an experimental way of learning about sharing space and the 'human reoccupation' of the city.

10 Cf Buster Simpson, 'Perpendicular and Parallel Streetscapes Stories', in *Belltown Paradise*, Chicago: Whitewall Editors, 2004, pp.41-60.
11 Ibid., p.42.

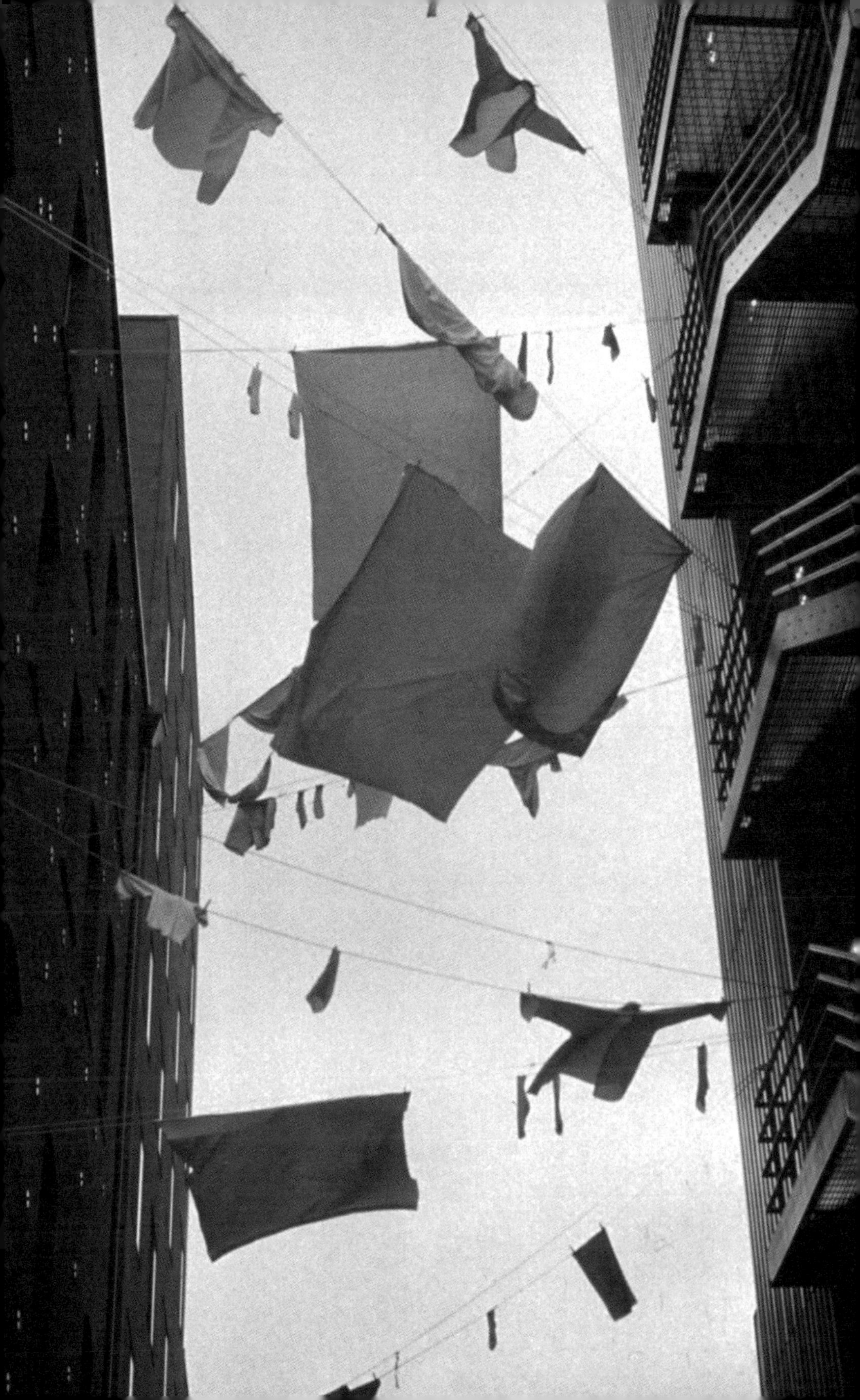

Another small insight was suggested by Marlène Puccini's art installation 'La tentation de voir' (the Sight temptation), realised in 1999 at La Belle de Mai in Marseilles, where the pristine white images produced digitally and the folded immateriality of the computer screens gave a strong sensation of love for the old women folding laundry in the open air environments of our minds and memories. As in her installation, the sight of the other in the city must stimulate the 'temptation', the desire to be in a place, and this presence should no longer have any 'function' apart from the pleasure to be nothing else than itself.[12]

Shared Solar: Shared clothesline between fixed income housing and condominiums, installation by Buster Simpson 1978.

[12] See Félix Guattari, *Cartographies schizoanalytiques*, Paris: Galilée, 1989.

Chapter 10

Mieke Schalk

Karen A. Franck

Måns Holst-Ekström

Gunilla Lindholm

URBAN CURATI
a critical practice towards greater 'connectedn

Han Meyer

Raoul Bunschoten

Chapter 10 Meike Schalk

Björn Hellström

NG

ess'

Doina Petrescu

There is one thing that makes curatorial practice very different from urban planning. *Curating* has come to mean making readings through exhibitions that in some way or another are cutting edge. The exhibition is by its very nature temporary, hence cutting. Urban planning is traditionally intended for an indefinite period of time. To operate with a model that allows for, or is, urban planning as temporary readings would be interesting, breaking up a traditional time–space relationship in the city. Is this the true temptation of 'curating'? Do we need urban planning? And what do we in fact mean by 'curating': is it the same as unplanning?[1]

The idea of 'urban curating'... seems to have possibilities to be a fresh way out from the conventional prisons of design styles or planning rules.[2]

Urban curating, a collaborative platform

The platform *urban curating* was launched in 2000 on the occasion of an invitation to a workshop in Jyväskylä, in central Finland. The organisers, the Alvar Aalto Academy, had selected twelve participants from eight different countries, and provided us with a brief on the issues of the post-industrial city. The task was to bring a concept to Jyväskylä that would serve as a starting point for continued work in small teams, and the aim was to develop a proposal for the city. I had never been to central Finland, and despite the broad information we had received from the organisers, such as maps, photos, and descriptions of structural problems, social polarisation, inaccessibility of the lakeside, etc., I found it difficult to form an image of the character of Jyväskylä for myself. It was announced that the workshop would be headed by Raoul Bunschoten, who had brought into play the concept of 'Urban Gallery', meaning 'a fluid form of public space that evolves in time, generating different definitions of public space and different ways of participating in it'.[3] In the programme, his role was defined as 'guest curator', while the participants

---

[1] Måns Holst-Ekström, art historian, urban curating roundtable discussion, 2000, see: *www.soc.nu/urbancurating/open/discussion.htm*

[2] Gunilla Lindholm, landscape architect, ibid.

[3] Raoul Bunschoten, 'CHORA Manifesto', *Daedalus*, 72 (1999): 51.

were expected 'to act as connectors of things and ideas, perhaps as curators'.[4]

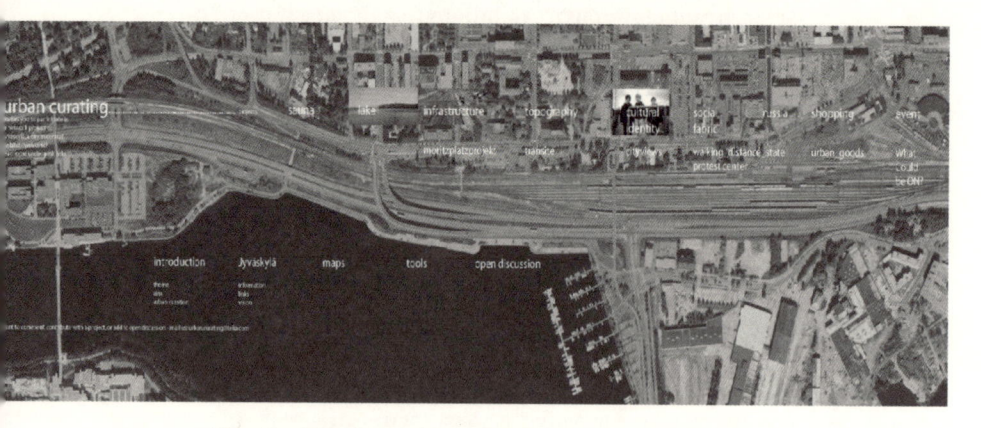

Helena Mattsson, Meike Schalk, Karin Willén,
urban curating Jyväskylä, 2000
(www.soc.nu/urbancurating/index/htm)

To prepare for the workshop, I joined forces in Stockholm with Karin Willén, an artist, and Helena Mattsson, an architect. We decided to take the role of the curator in the literal sense, and test the process of urban curating. For this purpose, a web-based platform called *urban curating* was released, for which we created a tool page, a roundtable page, and nine different themes to provide an initial form for contributions. Over one hundred architects, artists, landscape architects, theoreticians, and practitioners concerned with space were invited to participate in this network project by submitting comments, suggestions, references, models, prototypes, and concepts. We offered help in providing more specific information when needed, and supported the integration of the different contributions by leading a dialogue with the participants during the two-week-long process. It was our task to organise and edit, and to give the material the shape of a map or landscape that would guide the visitor through a web-based version of Jyväskylä, which represented not a physical proposal but an open platform for connecting different individuals and interest groups. This constituted our vision for the city at this stage.

4   From the second mailing of the organisers of
'Soundings for Architecture 2', Helsinki, 5 May 2000.

The project stayed at a discursive level, but a summary of it would read: what is missing in urban planning is a relationship of greater 'connectedness'. *Connectedness* in this respect could be described as a quality that can bring various aspects into the planning process that are otherwise not considered. For Karen A. Franck, connectedness represents a way out of 'unreflective categorization in urban theory, planning, and design', because it creates an opening towards 'the development of more inclusive, more complex, and more changeable categories'.[5] In practice, for her, it means a greater connectedness between different types of activities, closer spatial connection to support them, and a close relationship between the planner or designer and the client or the user. I think one of the most convincing recent examples of connectedness is what Helga Fassbinder was seeking in her 'Stadtforum Berlin' initiative after the fall of the Berlin Wall: an opening up to introduce different interest groups as actors to the negotiations in the planning process, and a new way of connecting them, to bring them together around one table. This aspect, which largely deals with our different personal relationships not only with our negotiation party, but also with the place under discussion, was strangely missing from Franck's list. But, I think, Franck would agree that we will not achieve better results through more or different rules and regulations, but rather through how we connect with the different actors and factors involved in a planning process. These are, on the one hand, the locality and everyone who is related to it from a position from within, and on the other, at a more abstract level, the planners, experts, and decision-makers who run the process from the outside. In addition, there is a wide range of different interest groups with communal, social, cultural, or economic agendas, each with a more or less personal connectedness to the actual planning case. Before we can even consider making our way through the forest of regulations, we have to establish these connections to start a planning process. The way we connect,

---

[5] Karen A. Franck deals with the notion of connectedness and inclusiveness in her essay 'A Feminist Approach to Architecture: Acknowledging Women's Ways of Knowing', in Ellen Perry Berkeley (ed.), *Architecture: A Place for Women*, London and Washington, DC: Smithsonian Institute Press, 1989, pp.201-16.

and to whom, will greatly influence emerging themes and tools for particular cases.

Camilla Holmgren, neighbourhood cinema
(kvartersbion), *urban curating* studio, 2001/2002

The practice of 'urban curating' was tested in a studio course conducted by Markus Aerni and myself at the School of Architecture, the Royal Institute of Technology, in Stockholm, during the academic year of 2001/2002. The studio *urban curating* was engaged in the question of the transformation from fluid movements to the formation of collective facilities. It examined the nature of existing institutions, and asked what future 'institutions' could mean. The notion of institution was considered, on the one hand, within a larger context, as an organisational body such as platforms, possibly as part of a global network, and on the other, as related to site, architecturally and physically, evolved from and directly acting within the local context. In this connection, different types of institutions were investigated, from the maximum institution of Sweden, to the most local collective facilities, such as a techno club, or a neighbourhood cinema. Following the observation that the meaning of institutions has changed in recent years, we were aiming at concepts that could take into account the process of transformation. Political institutions are becoming more dependent on economic institutions through privatisation and the liberalisation of markets. Strong resistance movements have visibly returned to the streets.

Simultaneously, globalised non-governmental organisations are moving into decision-making processes as important actors in world politics. The composition of actor networks has clearly changed in recent years. We have to respond to different forces, where new and old relationships are constantly redefined. As a tool, we installed and worked with the media of a website as a collaborative platform, whereby information, research, and findings from the fieldwork were shared among the members of the studio. The applied method was curatorial, meaning the role of the architect has shifted from the creator of objects to the mediator between actors, forces, processes and narratives.

Urban curating at Stockholm's harbour

As a consequence of our studio work, which we had discussed with the city planners and developers, we were invited to participate in a so-called 'parallel commission', together with five architecture offices. This is a common procedure within the planning process in Sweden, a sort of invited competition whereby the developer can pick ideas from different proposals.

Among the group of building professionals, our team, which included seven students from the previous studio, defined our practice as designing researchers, in this case 'institution'-based, and therefore capable of acting outside of the commercial context. As outsiders, we could take a critical position, and enjoy the freedom to find, invent, and test alternative tools. As we believe that the way information is collected will determine the entire planning and design process, we were eager to take other parameters into account beyond the standard templates that are usually superimposed on all sites in order to filter the 'right' information. We adapted a curatorial practice. What specifically did a curatorial practice mean for the site of Värtahamnen, a harbour area in transition? The actual situation of Stockholm's harbour is comparable to that of many other harbours around the world. The site conditions are similar to those we had encountered in Jyväskylä, an area in a liminal stage as well. The site is designated for change, and its potential is currently being re-evaluated. Many former activities have slowed down or disappeared already, though there are some exceptions, like the ferry traffic to Finland and the Baltic States, and a large business and conference hotel. The adjacent Frihamnen container harbour with its oil and gasoline storage is still served by a train that passes through the site of Värtahamnen, which itself has been cleared of oil tanks. Small-scale repair shops have popped up while some cultural institutions alternate with a few media companies. The remediation of the contaminated ground has turned a large part of the site into a field of excavation. The developer, together with the Stockholm City Planning Administration, called for the creation of office spaces for TIME-oriented (Telecom-Information-Media-Entertainment) companies, and envisioned a lively milieu, including cultural activities.

Andreas Ferm, Camilla Holmgren, Erik Hökby, Jani Kristoffersen, Anna Pang, Maria Sigeman, Lars Åstrand, *Urban Harbour (Storstadshamn; ett urban curating project)*, parallel commission entry, Stockholm, 2003

We started with a simulation of different ways of appropriating the site, and created various narrative strings of different agents, visitors, users, and actors who would meet there. Through this we learned about the potentials of the site. We collected our topics at the site, and found sound-identities, landscapes of different accessibilities, and variable programmes.

Then, we developed a toolbox of *instruments* and *prototypes*, which included loops (connectors), nodes (attractors), urban warehouses (variables), and acoustic reflectors and absorbers. The tools were used to review the site, and to redefine the relationship of city and harbour. In order to gain a better understanding of the area's history and its conflicts, we interviewed actors that were connected to the site, such as officials of the harbour authorities, and project leaders of the City Planning Administration. We involved experts of various fields in our research – including a sound engineer, a landscape architect, and the Interactive Institute in Stockholm – which allowed us to infuse other topics into an urban discourse, and to open up a broader discussion.

There is no standard procedure for a curator. The crucial difference from dominant planning practice lies in the gathering of urban knowledge that often remains unheard or is overlooked. This approach is based on the understanding that there are always multiple strands beside the official 'story', and that other and new perspectives must be found and produced. In this sense, urban curating is meant to work as a critical practice, a parallel and sometimes interfering movement to urban planning and design. It does not replace the traditional planning process, but it may trigger new modes of cooperation, interest groups, information and ideas. It therefore has the potential to consider the 'new', which differs from a traditional planning practice that mostly follows professional conventions, and orients itself on precedents, and often stereotypes. In this sense, with its way of organising and processing specific information, the practice of *urban curating* can fill the gap between problem and solution, and therefore serve as a complement to generalised and abstracted knowledge.

The sound of the harbour: reconsidering an urban 'problem'
Through the research material we received from different departments of the City Planning Administration, we learned that the noise of the harbour, especially the ferry traffic, was seen as an obstacle to developing housing here. According to the city planner, the best spatial solution to respond to this 'problem' was the city block, which created quiet interior courtyards. In contrast to this, during long discussions within our group, we reflected on the importance of sound as a particularly identity-creating urban feature. Further, sound is a factor that is experienced individually, which is 'learned'.

According to Pascal Amphoux, a French researcher who works on urban soundscapes, 'social and spatial practices today are interpreted solely from a visual point of view, however the auditory space can also contribute valuable information'.[6] The dominant attitude among politicians and specialists who work in the field of urban environmental research nowadays could be described as defensive. Basically, this attitude consisted of 'diagnosing the bad' in order to protect citizens from annoying sounds, supported by a static system of regulations whose goal is to normalise, regulate, and control, and which is implemented by building noise barriers, reducing traffic, erecting

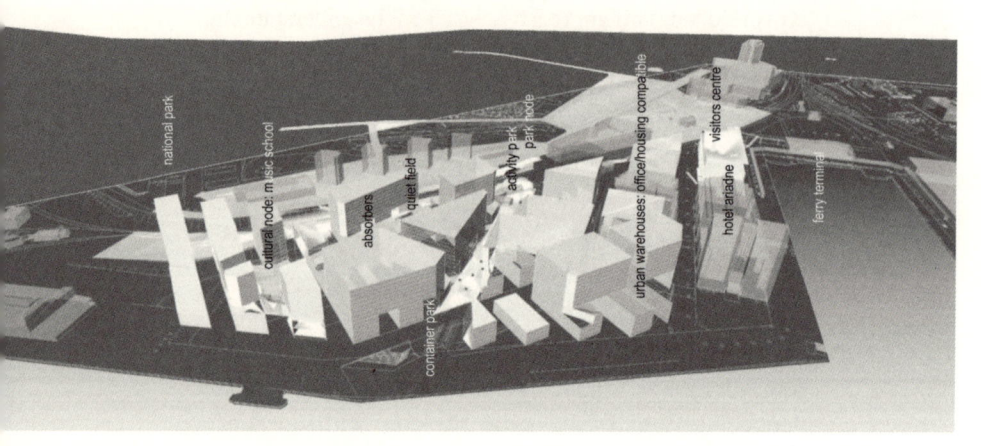

Urban Harbour, collective model

double walls, etc. Amphoux's work is based on the inverse attitude, 'to promote the favorable conditions of an actual and specific sound quality in urban space… In this sense a certain space that "sounds good" must not necessarily be quiet',[7] but as Amphoux states, should instead be regarded as a space where the sonic ambience seems to be adequate in relation to one's conceptions. He further suggests the practice of urban sound design. We arrived at questioning the erasure of the 'noise' of the harbour. Instead, we suggested actively experimenting with the physicality and materiality of the city fabric in order to manipulate noise, and to create an active soundscape as

6   Björn Hellström, 'Noise Design, Architectural Modelling and the Aesthetics of Urban Acoustic Space', unpublished draft of PhD thesis, Royal Institute of Technology, School of Architecture, Stockholm, 2003.
7   Ibid.

an informational landscape. Built structures, when considered in their properties as acoustic reflectors and absorbers, could potentially create more quiet areas as well as 'hot spots' of activities on the site.

Re-connecting harbour and city

Harbours/ports (doors) formerly represented the entrance to a city. Harbour cities emerged, and lived through their ports. What made the city was its port. This relationship has broken down due to the economical restructuring during the last decades, and large harbour areas have fallen into disuse and decay all around the world.

The planning process for Värtahamnen has currently stopped as a result of a veto by the harbour authorities in Stockholm. In this case, the harbour authorities, initially invited to take part in the planning process, were not 'interested' in participating. This is not unique – in most port cities, municipal and harbour authorities never enter into dialogue, because for them port activities and urban activities are mutually exclusive. There are a few exceptions. One of them is the case of Rotterdam, discussed by Han Meyer.[8] He sees the new design for the Kop van Zuid as an important sign: for the first time in decades, the city and the port have cooperated in a project, where port activities were no longer considered troublesome and inconvenient, but desirable and attractive elements of the cityscape. Crucial to this, Meyer points out, was the combination of uncertainty concerning the future use of port accommodation, and the desire for stability by designing public works in such a way that the city could benefit from them. In the near future, he expects to find a special mix of renewed old harbour buildings with new uses, and new harbour buildings with port facilities, and new buildings with urban facilities. As a still relevant example, Meyer refers to the historic case of the Commissioners' Plan for Manhattan of 1811, which expressed the idea of a complete equality between the city and the port, whereby the port

---

8   Han Meyer, 'The Port as Public Domain', in Aarie Graafland (ed.), *Cities in Transition*, Rotterdam: 010 Publishers, 2001, pp. 160-7.

economy was seen as part of the city, and was also integrated into the basic urban structure of the city. Here piers were formed as the natural continuation of the street system stretching into the water. In comparison, he points to Seattle as a successful contemporary example, where city and port are so interwoven that urban elements and port elements are interchangeable. As we can see, all three cases deal with an attempt to achieve greater connectedness through redefining continuity, by establishing a dialogue verbally and spatially, and through cooperation.

Official planning, developers' dreams, and 'other' practices
In the planning process, there is often not only a lack of communication between different actors but also little space for experimentation, and therefore few variations in design. In the Värtahamnen case, the planners embraced the late nineteenth-century city block as the solution to all their 'problems'. The city block reappears as an idealised urban typology that can now be applied to every site. This viewpoint reduces many of the possibilities, such as taking into account the differences between the inner city and the harbour environment, and working with the enormous scale, the fabulous view, and the public character of the site. In a developers' dream, 'urbanity' refers far more to a particular image of a city than to a certain idea of what urban life may imply. The glossy representations of a faked city often express this uncertainty, and, in the best case, they can be seen as pictures of hope for a brighter future mirrored in shiny glass facades, and smiling faces of well-dressed and happy pedestrians.[9]

We currently see a tendency whereby experimental approaches to urban questions emerge more often from the field of art, are funded by art and cultural institutions, and interestingly, but not surprisingly, are pursued predominantly

---

9   We need to add here that the main part of the 'developer's dream' is not so much to see the people happy but the financial profit of their development project. With a combination of cheap land and potential waterfronts, harbours are among the most desirable areas for corporate developers. That is why they are also the most sensitive and socially contested territories, constituting by their grand scale, the operation which will consequently lead to the final and irreversible stage within the gentrification of a city. From this point of view, to my knowledge, there has not been any socially successful model of harbour regeneration until now (editor's note).

by women practitioners from various professional backgrounds. I am thinking here of works that deal with different readings of a site, guided tours, the initiation of roundtable discussions that bring together officials and other parties, and the testing of new forms of participation and connectedness. On the other hand, many official planners, in general mostly male, also acknowledge the lack of situated knowledge in urban planning, but state they have no resources to integrate different viewpoints. Let us return to the figure of the curator here. As an independent cultural worker, (s)he escapes what we would call limitations, the 'myth' of the artist-architect whose authority is based on professional judgement and authorship, institutional bureaucracy that is founded on building law and abstracted knowledge, and the economic limits that are determined by the developer and the market. The curator moves in between these features. Official instruments for urban planning are strict, and not especially flexible. There is a need for interference, which may take different faces, and which must be constantly redefined. We felt strongly that a curatorial practice could make a difference here.

Chapter 11

muf

Kim Trogal

S. Crow, A. Brown, S. Essex, H. Thomas and C. Yewlett

OPEN KITCHEN OR 'COOKERY ARCHIT

Prue Chiles

Katherine Shonfield

Gillian Rose

Michel de Certeau

Laura Trippi

Henri Lefebvre

Chapter 11   **Kim Trogal**

Jeremy Till

ECTURE

Jennifer Bloomer

Luce Girard

Raoul Bunschoten

Gilles Deleuze and Félix Guattari

OPEN KITCHEN OR 'COOKERY ARCHITECTURE'

The Open Kitchen is a theoretical and unrealised proposition to engage a group of women in urban regeneration right from the stages of planning down to detailed construction. The proposition outlines a methodology for participation in urban regeneration through socialising, cooking, eating and building where the domestic kitchen and cookery are taken as a paradigm for architecture and set as a hypothesis for a project.

Domestic cookery is traditionally women's everyday practice of making in the private sphere, which, taken as a theory, offers a model for architecture that is critical yet optimistic. The instant attraction of taking cookery as a concept is the respect and celebration of difference and diversity we enjoy with it. There is a shared social experience of making and eating food that is not paralleled in the building industry. Throughout the project I was looking to develop a deeper understanding of culinary practices, to understand their logic and qualities and to investigate how these could be transposed to a practice of architecture.

The Open Kitchen was developed with Doina Petrescu and Aidan Hoggard as my thesis project for postgraduate diploma in architecture at Sheffield University. The project sat within the wider agenda of the design studio entitled 'strategies of transition'. In the context of the studio we were encouraged to take a critical position with regard to the role of both architecture and the architect. We were challenged to understand and develop an architecture whose role was not just a building but something that could become a tool within the transformation of social, political and economic conditions. We acknowledged that a 'neutral' position is an impossible position and were asked to develop a platform from which we as individuals could be clear in our roles; to understand how our own actions as architects are political, in the broadest sense. Although the Open Kitchen is a theoretical project, it has a specific spatial and political situation. It has its roots in real life, real conditions, places and people. It is situated in space and time.

The 'kitchen' as an alternative catalyst for development

The Open Kitchen is sited in Foxhill, an area on the periphery of North Sheffield, between the city and the Peak District and on the cusp of an urban/rural edge. As the area stands now, it consists predominantly of low-rise council housing and is one of the poorest areas in Europe. It is on the margins geographically, economically and socially.

The typical approach to regeneration in Sheffield in the past has been one of *tabula rasa*; the very approach that created the area of Foxhill during the city's post-war 'slum' clearance programme. From the late 1950s to the early 1960s, inner-city residents were re-housed out of the city in a clean but dismal suburbia, without any of the aspects of a suburban lifestyle that might have made it a desirable place to live.

More recently, Sheffield city centre has received huge investment and has undergone massive physical transformation: cranes across the skyline, new hotels, apartments, galleries and shops. The principally economically driven development in the centre risks the displacement of a population who cannot afford to continue to live in the area, and their exclusion from the new 'public' facilities.

At the time of developing this project, the council proposed to sell parcels of land in Foxhill to a developer for the construction of private luxury apartments. The approach to regeneration in this case is a similar strategy to that adopted in the city centre: the hope that knocking council houses down and introducing private investment will bring the middle classes and their money into the area. Instead of acting as a catalyst for regeneration, this approach promotes a process of gentrification. The risk of marginalising the residents of Foxhill through increased privatised provision of facilities is high. Although the developments may promise increased employment opportunities, these opportunities are limited in their scope. There would be little social benefit for those who cannot afford.[1]

---

[1] The employment generated by the construction process itself would be awarded to large contractor, who may employ some local labour, but on the whole would be unlikely to support small local business in the process. The small-scale and longer-term employment that would have been generated included the provision of services and amenities to support the development, such as small shops, supermarkets, and within new private facilities for child care and sports and leisure facilities. This, of course, brings benefits, for instance, the chance to work, get experience, get money, get a reference. However, this economic regeneration centres around attracting wealth and maintaining a status quo in the distribution of wealth. It does not transform existing economic structures, or empower people to do so by themselves.

For more empirical research on the effects of gentrification in different contexts, see, for instance, R. Atkinson and G. Bridge, *Gentrification in a Global Perspective*, London: Routledge, 2005.

Open Kitchen or 'cookery architecture'

The Open Kitchen focuses on a specific site earmarked for private development and offers a counter-proposal. The site was then occupied by three blocks of council-owned maisonettes, a few months prior to their demolition. They existed in a bizarre in-between state of partial habitation; many of the flats were empty and boarded up after the families living there had been relocated. One of the maisonettes was inhabited by a Sure Start[2] drop-in centre. The centre played an important local role, socially and educationally, bringing young mothers and their children together and giving them the opportunity to talk, and get advice from each other as well as health professionals.

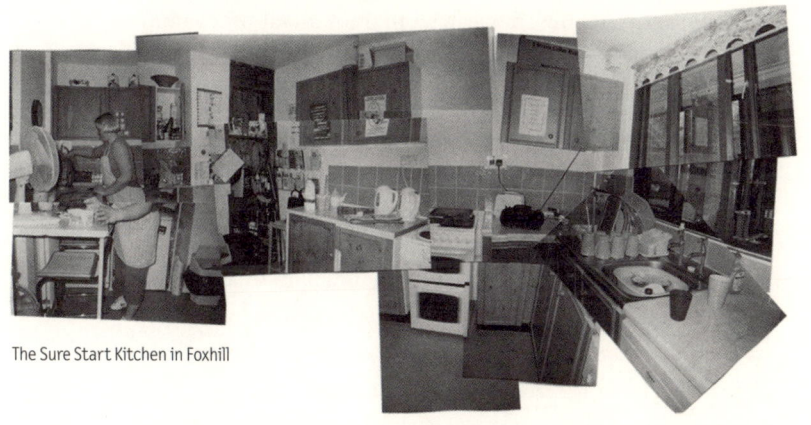

The Sure Start Kitchen in Foxhill

The Sure Start centre had built up a network of individuals in the area over the previous three years and although the building they inhabited was less than ideal, staying was preferable to being relocated, as they feared losing the relationships they had worked hard to build up.

I spent time in the kitchen of the flat, interviewing some of the women there. The kitchen already existed as a place of exchange and communication; it was already a social space and a space of making. I began to think about this place as an alternative catalyst for development.

The role of the kitchen in this context had already been expanded to have a wider local significance. The informal environment is a place for exchanging gossip, ideas, advice,

---

[2] Sure Start is a government programme offering health, education and child care support to families with children under the age of four. See *www.surestart.gov.uk*

cooking tips and food. The Open Kitchen proposes a further extension of this exchange, becoming a base for consultation and dialogue on design as well as a place for making urban/architectural products. The kitchen becomes a place for discussion and sharing ideas on wider urban and public issues.

The interviews signified the beginning of the project; they were the next step in expanding the role of the kitchen.

Involvement of 'others': scenarios and time

The studio used 'scenario making' as a tool to develop our proposals. The scenario represents both the process and the parameters of the project over time. The scenario as a tool allows an architect or designer to think about the evolution of a project and to actively acknowledge and encourage other events and people to enter the process. A scenario allows an architect or a designer to take another position in this process. Within this context, Raoul Bunschoten talks of planners as urban animators and curators:

> Cities need animators that link, weave, stir, stimulate. Cities need 'bohemian' planners for the next century, planners who are autonomous, able to grasp multiple aspects of reality and refashion them as narratives to link an increasingly heterogeneous population in decision making processes. Planners have to include festivals, collect dreams, conflicts, mobile phone technology, smart materials and new hybrid cultures in their urban designs, as well as the more conventional planning of infrastructures, housing and other physical structures. The urban animator has to communicate reinvented spaces and solicit participation and complicity.[3]

The scenario can represent possible relationships between the parameters involved in a project: the social, financial, political and personal, the 'multiple aspects of reality', and can allow other events and cultures to enter into urban design. In the Open Kitchen, the scenario is used to investigate how a self-build programme for the area might be initiated by developing

---

3   Raoul Bunschoten, *Urban Flotsam*, Rotterdam: 010 Publishers, 2001, p.41.

a culture of exchange between the university and the women at the Sure Start centre. The scenario, like a recipe, outlines the possibility of making something, it details things that are needed such as people, places, resources (the ingredients) and suggests how they can mix together and when.

The scenario begins with a 'meeting' in the Sure Start kitchen. The kitchen, already a place for chatting, sharing and making, is expanded/extended/inflated over time to become the 'urban kitchen': a paradigmatic extension, a 'para-site' to the existing kitchen. This is a place for discussion and sharing ideas about building. In this scenario, the kitchen forms the basis for initiating dialogue about buildings as well as the base for actually making buildings in a self-build programme.

'Performed connection' and 'feminine' production of space

Whatever we are making, food or buildings, the process can become a social event from which a community can develop through performance.

In her analysis of community arts projects in Edinburgh, Gillian Rose emphasises the importance of the performative nature of the projects and events. She highlights the importance of the act of 'making the object together' rather than just the object itself. She argues that participation in community arts is a dynamic process, where the project is 'not *always* the expression of an anterior identity but can be understood as a development of identities',[4] that is to say, the participants' shared involvement and experience change both them and the outcome of the project. She terms this dynamic condition '*community of performed connection*'.[5]

In the context of the Open Kitchen, the act of making the objects involves a long process between students, local women, those in the council, the local regeneration agency, other

---

4   Gillian Rose, 'Performing Inoperative Community: The Space and Resistance of Some Community Arts Projects', in Steve Pile and Michael Keith (eds), *Geographies of Resistance*, London: Routledge, 1997, pp.184-202.

5   Ibid.

Cake workshop, 6 March 2002

builders, local manufacturers and specialists. This process involves continual learning, experimenting, talking, designing and making on site, through workshops, discussions and doing. For each individual, it demands working with others whom you would not normally encounter and undertaking work that you would not normally do; it is a process that could at once be liberating and frustrating.

This process is made up of collective and shared experiences. The identities of the project and those involved are continually transformed and simultaneously transform the process itself. It is not planned from the start, but dependent on the interests and desires of those involved. Once participants develop courage through their initial involvement, it is imagined that the process will take on a life or identity of its own, one that is interconnected with the identities of the participants.

Experiences such as the social events of cooking and eating together can constitute a very personal exchange. As a paradigm, this more intimate situation has the potential to transform everyday situations within the design and construction process to create a 'community of performed connection'. The Open Kitchen scenario involves the transformation of these situations. Planning meetings are taken out of their context and city councillors are invited to come to the (urban) kitchen instead, to cook and eat with local people: a new type of meeting is created. The consultation event is not 'over lunch' but *is* lunch. A meeting can become a more informal and sociable event, one that is less institutionalised than a meeting in the council chambers. Such a meeting is both a space and a social product initiated by a 'feminine' performance of informal exchange.

I began to think about food as a way of initiating dialogue and so I invented a game using cake 'models' of the three buildings scheduled for demolition. In a workshop with first year architecture students, we took the cakes up to the Sure

Start centre and joined a social morning. We invited the group to slice, ice, add, rename and eat the models, while generally discussing their feelings about the area as we did so. The game was intended to be an informal and funny way to initiate conversation about the area, discussing both problems and ideas. The approach allowed us to talk about buildings and facilities in an abstract enough way to begin to consider ideas and possibilities, not just the stark reality of the area. In addition, we were not talking about architecture, about beautiful new buildings, we were talking *around bits of cake*.

The ideas that emerged from the workshop were taken as a programmatic basis for the scenario involving the strategic habitation and transformation of the semi-vacant structures. The scenario proposes that the ideas identified in the workshop develop over time as series of small self-build projects. The workshop itself signifies another step in the expansion of kitchen.

Everyday practice as design tool

Cooking is women's everyday practice in the private sphere and as a paradigm it can allow new definitions and methods to emerge within a participative approach to architecture and planning. De Certeau describes the political significance of 'everyday practice':

an everyday practice opens up a unique space within an imposed order... [it] is relative to the power relations that structure the social field as well as the field of knowledge... to appropriate information for oneself... is to take power over a certain knowledge and thereby overturn the imposing power of the ready made and pre-organised... everyday practice patiently and tenaciously restores a space for play, an interval of freedom.[6]

Everyday acts in the kitchen, even washing up, can have poetic resonance and create a social space around the performance.[7]

---

[6] Michel de Certeau, Luce Girard and Pierre Mayol, *The Practice of Everyday Life*, Vol. 2: *Living and Cooking*, trans. Timothy J. Tomasik, Minneapolis, MN: University of Minnesota Press, 1998. pp.254-5.

[7] Ben Kinmott, in one of his art projects, uses washing up as a generator of social situations. See Laura Trippi, 'Untitled Artists: Projects by Janine Antoni, Ben Kinmott, Rirkrit Tiravanija', in Ron Scapp and Brian Seitz (eds), *Eating Culture*, Albany, NY: State University of New York, 1998, p.138.

Washing up is a part of the process of making in the kitchen and so, in the Open Kitchen, places like the sink find their equivalent on site. The small-scale transformations that take place in the kitchen – expanding, evaporating, removing, pouring, cutting, spreading, mixing, places of cleaning, storage, waste – find their equivalents on site, their locations and natures changing over time as the scenario develops.

In this case, metonymy [8] is used as an allegorical tool, where the attributes of something (cookery) are transposed to become emblematic of the project and a process for architecture. Jennifer Bloomer locates allegory:

as the place where ideas repressed by the dominant may be hidden, as well as the place where the personal and the political may coincide...

In allegory 'any person, any object can mean absolutely anything else'... at the same time, these things that are used to signify acquire a power that makes them then different, on a higher plane. [9]

Metonymy allows new understandings and meanings to evolve. As such, the paradigm of cookery provides a critique and possible opportunity for planning and architecture. First, is the question of difference, different tastes, different wills: how can a culture of difference thrive in the production of architecture as it does in cookery? Then, appropriation and personalisation: how can participative architecture be truly that, where design consultation is not coercion but allows people to take ownership, to appropriate something according to their own desires, yet still operate within a collective authorship? And the freedom of experimentation: how can people find the freedom to experiment with building and construction in the same way we do when cooking? How can architecture (and architects) be less controlling and

---

8 Metonym stems from the Greek *metonymia: meta*, indicating 'change' and *onyma/onoma*, a 'name'. It is used in a transferred sense, referring to an attribute or something emblematic (for instance the crown referring to the monarchy) rather than a synecdoche that takes the part to refer to the whole.

9 Jennifer Bloomer, *Architecture and the Text: The (S)cripts of Joyce and Piranesi*, New Haven, CT: Yale University Press, 1993, pp. 49–51, referring to Frederic Jameson, *Third-World Literature in the Era of Municipal Capitalism*, and also quoting Walter Benjamin, *The Origin of German Tragic Drama*.

proscriptive in the process of making a project to encourage the creativity of others?

'Appetite': a logic of taste, pleasure, difference and participation

Issues of difference rarely feature in conventional planning and consultation. The British Development Plan system has received much criticism over recent years,[10] particularly in regard to public participation and consultation.[11] The problem with this system is that it does not easily solicit participation. Public exhibitions fail to attract public interest, as people often cannot spare the time or effort. The development plan system that seeks to engage some kind of 'collective will' in fact only involves the limited interests of various developers, interest groups and organisations. This method that seeks to gain a consensus on the plan actually achieves the opposite.

One of the attractions of taking cookery as a paradigm is the enjoyment of difference, diversity and of taste we take from it. Within a traditional system of planning, advisory groups are set up to run consultations and attempt to find 'common ground' between participants, where there is little or no flexibility or space for difference.

From participation in planning to building, the traditional role of the architect (as author) is jeopardised in an attempt to create a more democratic process. The idea that participation

---

10 See S. Crow, A. Brown, S. Essex, H. Thomas and C. Yewlett (eds), *Slimmer and Swifter: A Critical Examination of District Wide Local Plans and UDPs: A Research Report Commissioned from University of Wales Cardiff*, London: Royal Town Planning Institute, 1997.

11 The UDP system is in the process of being replaced by Area Action Plans (AAP) where public consultation is mandatory. At the time of writing this project, the UDP operated broadly along the following stages:
 1 Unitary Development Plan (UDP) prepared and put on display.
 2 If there are any objections, these are considered.
 3 A revised version of the plan is prepared and put on display once more.
 4 If there are any objections (you may only object to the plan if you did so on the previous occasion), a revised plan is prepared once more before going to inquiry.
 5 The planning inspector prepares a report, and...
 6 If no further modifications are made the plan is adopted.

Although the AAP is an improvement on this in that it attempts to make more opportunities for people to find out and speak out about development, as a system, it still suffers from similar problems. The process of consultation is structured around gaining feedback on proposed plans rather than a creative engagement with the transformation. It is a system of planning aimed at building consensus. For an analysis of consensual approaches, see Tim Richardson and Stephen Connelly, 'Reinventing Public Participation: Planning in the Age of Consensus', in Jeremy Till, Doina Petrescu and Peter Blundell Jones (eds), *Architecture and Participation*, Abingdon: Spon Press, 2005, pp. 77–99.

#04:
Kerry mentions the lack of clean play facilities for her daughter as the nearest parks in Foxhill and Parson Cross are used by drug users who leave needles, bongs, foil etc. on the floor.

Kim: what about putting a playground on the roof, somewhere for your children to play that's accessible from this maisonette, like a garden or something?

Amy, ices a slide and a roundabout onto the roof.

[...] Kerry: If you could come up with something that would solve this area...

Kim: well... I'm not sure there's just one thing that would do that, on it's own I mean... in one go...

Kerry: no...

Kim: maybe lots of little things?

Kerry: maybe... [definately still unconvinced about keeping the maisonettes and what might be possible]

#04:
I suggest creating a roof garden/ play space on top of the maisonette we are in.

SURE START PLAYGROUND

#10
Kim: maybe we could have a kind of cafe and incorporate that healthy eating thing you mentioned?

Kerry: it was meant to be kind of healthy eating based on a budget

Kev cuts the cake to create a terrace for a cafe that could become part of this programme.

Kim: have you just cut a hole in our playground?!

Kev: oh yeah... erm...

(more giggles)

#10
CAFE / RESTAURANT "HEALTHY EATING ON A BUDGET."
We open a restaurant to encourage healthy eating.

Kerry: we could also have a big open recreational area where all the nutcases can go to drive their cars at 100 mile an hour as well and burn round corners. (#11)

Kim: like a kind of ralley track?

Kev: do we need a building for a ralley centre [is poised with a knife over the carrot cake]

Kim: I don't know...

Tam: I think if they had somewhere to drive their cars they'd be quite happy. [...] It would be good if they were going to do it.

Kerry: yeah, to me that's what it needs... but if you spoke to the lady next door she'd probably tell you something completely different. She swapped her house on... somewhere near Norwood unofficially to get back here on these maisonettes becuase her kids... well its alright me saying there's all this trouble and that but her kids... she's had less trouble they've achieved better at school and everything since she's moved back here. So it's not all negative.

Kim: do you think she'd still mind living here if there was all this going on?

Kerry: She'd be grateful of that [points to rehab centre and gym]

Kim: If she has children she might not like your coffee shop!

Kerry: no!

Tam: but why? it's only ganja isn't it? They wouldn't be shooting up.

Kerry: at this meeting they said that 50% of housing in Parson Cross wants re-roofing in the next five years and because they're not renting these two bedrooms they're not getting the money to do the repairs they want to level them and build a better standard of housing. They think that if they have a better standard of housing they, and hopefully attract well he didn't say it but a better class of people but that's what he meant. I felt like saying just that just because they've got a better education that they're not going to smoke draw then. Just because they come from Crosspool or something, do you think its a problem in Foxhill and Parson Cross, know what I mean. They're not going to stop doing that and its all right getting different people in.

Tam: yeah, you can't just shove it under the carpet, there's still going to be the same problems.

Kerry: So if you can give 'em somewhere to race their cars and go mad and do it with crash abrriers and all that and give 'em somewhere to smoke draw let them do that... do you know what I mean?

#01
Kim: I'm going to extend the sure start maisonette, make it bigger and a place where we could do this for real, with real bits of building... instead of bits of cake... what do you think of that? would you do it real?

Kerry: erm...[shakes head] not in the maisonettes, I personally wouldn't the only thing I'd do with these maisonettes is I'd flog 'em off and make 'em private...

Tam: they'd be quite nice then, becuase they vet the tennants don't they?

[...] Kerry:If they were young proffessionals or something, they'd be much nicer. Because those stairs up there are absolutely disgusting, I wouldn't even walk down them...

Tam: have they got a lift?

Kerry: no they haven't, try and get a double buggy down there. If you made them private, people couldn't wee in it or draw in it

#01:
I place my urban kitchen outside the sure start maisonette.

KIM'S KITCHEN.

#02:
Amy sugg
but after c
decide on

#11
Kerry wants somewhere for young kids to race cars around so they don't do it on her road.

RALLEY TRACK FOR RACING CARS.

Open Kitchen or 'cookery architecture'   179

ainment
sm we

Kerry's Hash House

s a coffee house like in Amster-
of confronting what happens
ather than ignoring it.

#02
#06
#08
#05
#09 #07

Tammy's Creche

#02
Tam: The main thing is why it's like it is round here, why there is alot of problems and alot of trouble is beacuse there's nothing for the kids to do.

Kerry: no, and nothing for adults at night

Tam: they're causing trouble 'cause they're bored, vandalising, smashing car windows in

Amy: maybe we could put a little cinema here, round here?

Kim: go on then, here's some cake...

Kerry: [...] they wouldn't build a cinema round here, they'd nick the camera

Tam: they'd just trash it... a little arcade they'd probably use more, with pool tables in it, pin ball machines that sort of thing... they'd use that wouldn't they? it would be like a gathering place.

Kerry: but then it'd get shut down 'cause they'd all be smoking draw in it. (#03)

Kim: aren't you the pessimist?

(giggles)

Kerry: but no, that's how it is

Matt turns off the video as we discuss drugs- Tam wanted this spoken in confidence.

We place a 'coffee house' by the maisonettes- I infered from this part of our conversation that Kerry and Tam felt that the local authorities ignored drugs issues (of all classes) and that they pretended it didn't happen. We speculate about whether relaxing cannabis laws in this area might make the area more relaxed (with regard to vandalism).

Tam: have you got to go back and tell your tutors 'this is what they want', a hash house! (giggles) 'we went to this playgroup and this is what they want!' (giggles)

#05:
REHAB CLINIC & DAYCARE CENTRE.
Kerry wants to see some rehabilitation facilties for herion addicts in the area. Amy turns the middle maisonette into a rehab centre.

#05:
Kerry: did you see in The Star this week, a big supplement in the middle, something like 'drugs war in sheffield'?

Kim: is there a drugs war in Sheffield?

Kerry: yeah well they're clamping down on everybody, there was a big bust in Doncaster, [...] I've got a passion about this I've got a friend whose daughter is a heroin addict and its a nightmare and she can't get help, she can't get this and she can't get that. She's classed as an emergency case and she's on a six month waiting list, know what I mean?

#06:
KERRY'S FAMILY PUB → things for children
Kerry wants a family pub, where she can have a drink and there are things for her daughter to do.

#06 [...]Kerry: you know somewhere where you can take your kids, a family kind of pub [...]

Tam: a creche would be a good idea, a nursery and a creche. (#07)

Kerry: a creche more than a nursery as there's "little rascals" on Wolfe Road, but you know like a shoppers creche where you can book 'em in for two hours in the morning so you can go to the gym or whatever.

#07:
Tam wants to go to the gym more and so would like somewhere nearby to leave her son for an hour.

#08 + #09 Amy: we could turn this block into a gym?

kerry: with a swimming pool attached because that's the problem. There's a good gym down the road that I go to but the problem is that there's no pool...

kim: How big are swimming pools?

Kev: quite big?

Tam: 25 metres!

(giggles)

Tam: we don't want an olympic one!

Kerry: yes, there could be a gym with a pool, and shower block kind of here (points) to the model and we could incorporate this bit of land here and have a football pitch.

With blue icing, amy ices a swimming pool

Kim: the pool could be half inside and half outside,

Kerry: yes, indoor and outdoor..

[...] Tam: it would be great if it would [happen].

#08 + #09
Tam is keen to get fit again after having her son, Amy suggests using the end maisonette for sports facilities.

TAM & KERRY'S GYM + SWIMMING POOL

INDOOR + OUTDOOR POOL

or 'community architecture' is by virtue more democratic is not necessarily true. The process establishes a power structure between the expert knowledge of the professional and the tacit knowledge of the layperson. Jeremy Till writes:

> a power structure in which the expert architect assumes authority over the inexpert layperson... is unacceptable if one aspires to a participatory process that empowers the user... participation cannot be achieved through the disavowal of expert knowledge. Nor is the solution to make the architect more accountable by making it more transparent... Instead, a move towards transformative participation demands a reformulation of expert knowledge and the way it may be enacted.[12]

The role of the architect must be carefully considered, a role that moves from author to facilitator, using and disseminating their knowledge in a way that participants can work with them creatively. We are all active participants in the process.

The refiguring of the relationship between architecture and 'user' as a political act highlights the importance of consultation as a creative process. It can provide the opportunity for exchange and space for difference. In this regard, the Open Kitchen employs the metonymic power of 'appetite'. 'Appetite' can define the programme and the brief of a project in terms of 'what we want to eat', what we desire, all dependent on our individual tastes. With 'appetite', each person has their own taste; their own appetites within the act of cooking or making together. There is no singular collective will but multiple different appetites within a collective; there is some common ground but emphasis is on a 'split', a difference in taste.

With 'appetite', the emphasis is on difference and its logic is applied to the Open Kitchen at various stages. It has implications for consultation and participatory practices as well as identifying one of the key problems in consultation, namely 'motivation'. How to keep people interested, engaged and motivated to see a project through? How to keep momentum going? What gives them appetite for a project?

---

[12] Jeremy Till, 'The Negotiation of Hope', in Peter Blundell Jones, Jeremy Till and Doina Petrescu (eds), *Architecture and Participation*, Abingdon: Spon Press, 2005, pp.30-1.

'Appetite' brings a logic of taste *and* pleasure. Following this logic, the strategic development of a project is altered, namely that the most appetising part of a project is built first.[13] This relates to the expansion of the kitchen and the creation of a public café as a workshop for making food and buildings. This is the place that provides the potential for future possibilities, a small-scale construction of the strategy. Katherine Shonfield describes the relationship between the small scale (the particular) and the strategy (the general) in the work of muf:

How do you develop a city-wide strategy when you are fascinated by the detail of things? And how can you make something small scale in the here and now if you are driven by the urge to formulate strategic proposals for the future?... urban strategists move from the general themes to particular instances, muf's work, on the other hand, develops the particular to the general and back to the particular... it can be described in abstraction:

1 The close interrogation of the up close and personal (detail)
2 The extraction of what the personal can tell you about the general (strategy)
3 The reformulation of the strategy in the here and now: a small scale construction of a future 'what if...' (detail)'[14]

In this scenario, the café is 'a reformulation of the strategy'. It is a place to meet, chat and think about future projects, and expands the existing kitchen – both physical and metaphorical extension. It is a small-scale construction of a future 'what if...', it encourages thinking about possibilities, it is a flavour of future change. In this way, the Open Kitchen is the antithesis of a monumental building. According to Lefebvre:

monumental buildings mark the will to power and the arbitrariness of power beneath signs and surfaces which claim to express collective will and collective thought. In the process, such signs and surfaces also manage to conjure away both possibility and time.[15]

---

13 This is derived from muf's 'Pleasure Principle'. See muf, *This Is What We Do*, London: Ellipsis, 2001, p.17.
14 Katherine Shonfield, ibid., pp.14-15.
15 Henri Lefebvre, in K. Michael Hays, *Architectural Theory Since 1968*, Cambridge, MA: MIT Press, 1998, p.184.

The scenario of the Open Kitchen does not claim to express 'collective will' and 'collective thought' but could hopefully exist as a product of multiple wills within a 'collective'. It attempts to encourage future possibilities over time through small-scale constructions of the strategy in the 'here and now', rather than 'conjure away both possibility and time'.

The language of cookery: appropriation and experimentation through recipes
As 'appetite' emphasises difference and desire, so can the language of cookery. In particular, the recipe can encourage variety and difference as well as personalisation and appropriation. In the scenario, the 'recipe' is proposed as a tool in architectural processes.

Like all human action, these female tasks are a product of a cultural order: from one society to another, their internal hierarchy and processes differ; from one generation to the next in the same society and from one social class to another, the techniques that preside over these tasks, like rules of action and models of behaviour that touch on them, are transformed. In a sense, each operator can create her own style according to how she accents a certain element of practice, how she applies herself to one or another, how she creates her personal way of navigating through accepted, allowed and ready-made techniques.[16]

Each person creates their own way of cooking, their own way of 'navigating through accepted, allowed and ready-made techniques' and the recipe is something that can encourage this. In investigating the parallel between appropriation in cookery and appropriation in design, I proposed a double analogy. I wrote one recipe for site-mixed concrete and re-wrote that section of the national building specification to read as directions for baking a cake. Through this I aimed to highlight that the national building specification, due to its rigidity and proscriptive nature, as well as highly specialised

---

[16] Luce Girard, 'The Nourishing Arts', in Michel de Certeau, Luce Girard and Pierre Mayol (eds), *The Practice of Everyday Life: Vol. 2: Living and Cooking*, trans. Timothy J. Tomasik, Minneapolis, MN: University of Minnesota Press, 1998, p.156.

Open Kitchen or 'cookery architecture'   183

Cooking-up construction

Introduction:
Following on from a short workshop on edge conditions, we focused on a specific notion of linguistic edges, looking at the way language creates an edge or boundary between the 'inside' and 'outside' of architecture. The use of language is an integral part of architecture, using many different vocabularies: theoretical, managerial and technical, to communicate with different parties involved in the building process. These languages have developed over time and have become abstract and specialised. Architectural language also excludes lay people to preserve the status of the profession (Jeremy). If this language doesn't mean anything to anyone outside the industry is it possible that we are in danger of reducing the value of architecture to other people. Arguably, this language also in part defines the profession by taking focus off other issues, for example architects become absorbed in the act of building or construction rather than questioning.

Abstracted language is prevalent in the 'building specification', the naming of parts for construction and the way in which they are put together. These are specific instructions from the architect to the contractor and are legally binding. They are explicit documents, supposedly easily understood and acted upon. In theory these documents are a standard that the building industry works to and enable the architect to specify something twice in separate projects to achieve an identical outcome. In reality, this is largely an unquestioned pretence.

The language of 'building specification' helps to maintain the current pattern of building, where the client / user is generally excluded from the process. The constructional part of the process is hidden from them and the part they are most alienated from. Hence many people feel that building is something they do not know about and cannot contribute to.

Proposal:
Using the language of food preparation and cooking we hope to make a critique of the language of building specification, potentially making this part of the building process more accessible (although not necessarily). Food is a basic necessity that everyone is familiar with. People enjoy experimenting with food and are not afraid to question or criticise it. Although we realise that the food industry has its own complexities, we feel that the language of cookery is an interesting and appropriate medium through which to investigate architectural language. Like the construction of a building a meal is something that many people can contribute to but unlike the majority in the building industry it is something that celebrates cultural variety and promotes difference.

Site-mixed concrete

I first came across this dish while visiting a vegetarian friend in Sheffield. Varying from one region to the next, this recipe can take boring old concrete in a completely different direction. Best served on a bed of rusted reinforcement bars, this dish will provide a delicious base for any meal. The smokiness of Portland cement, the fragrance of freshly ground Italian marble coupled with the classic taste of sharp sand impress your friends with any number of ad-mixtures that can transform the texture, colour and freshness.

One mixed batch serves 3m2 at 150mm thickness

Preparation time: approximately 7 days

Ingredients:
Cement, 5 heaped shovels
Aggregate, 5 shv
Sand, 5 shv
Water, one bucket
Ad-mixtures (to taste) — 2 pinches of plasticiser

Firstly, ensure that your larder is dry before storing ingredients, and make sure that the sand and aggregate are kept clean and dry, not forgetting to store each type of cement separately. Prepare suitable sturdy container or mould making sure it will withstand the weight of the concrete. Meanwhile, spoon cement, aggregates, sand and water into the mixing bowl and season with ad-mixture to taste. Stir thoroughly until it forms a gloopy consistency. When ready, gently pour the mixture into mould remembering to keep freshly poured concrete moist at all times by covering with damp cloth. We recommend that the concrete be set to one side for at least seven days. Don't be tempted to check setting of dish as this will spoil the surface of the mixture. Gently remove the mould, taking care not to damage the concrete. To finish, polish till golden and serve.

Note: If seasoned with additives, setting time can vary.

Henry's favourite!

Approach One. A Recipe for Site-Mixed Concrete.

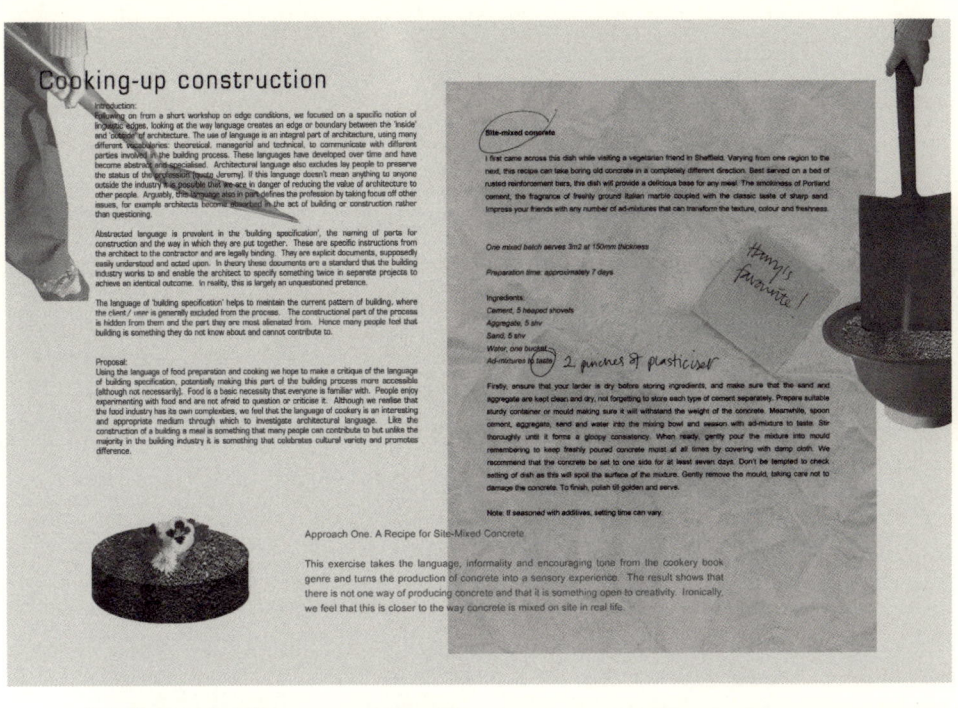

This exercise takes the language, informality and encouraging tone from the cookery book genre and turns the production of concrete into a sensory experience. The result shows that there is not one way of producing concrete and that it is something open to creativity. Ironically, we feel that this is closer to the way concrete is mixed on site in real life.

Recipe for concrete making

Concrete making in the kitchen

language, was unsuitable for a more participative approach to design and construction, particularly self-build. The specification is a legal document between architect and contractor, not architect and 'user-builder' in a self-build context.

Through the recipe I aimed to demonstrate that concrete could become a more sensitive material by allowing people to contribute or even create their own process. This material sensitivity could be achieved by the addition of very small things embedded into the surface, additives to the concrete, altering the ingredients: aggregate, formwork, and so on. The concrete can become a register of who made it and when: it is a product of what goes into it, the temperatures and conditions it is kept in; it keeps the traces of its production.

Concrete samples

The recipe is an informal piece of writing that can encourage experimentation. In testing the concrete recipe, I used everyday items as formwork and experimented with different ingredients to be added. It was trial and error. I had no idea how it would turn out.

There is a logic of experimentation and uncertainty to cooking; there are no definite plans as to how it will all turn out. You cannot 'know' everything and control everything 'rationally'. In the kitchen we are working with accidents and chance but guided by our senses, taste, touch and smell. Our gestures are small and intimate; they are of the body.

How might this different logic affect how we as architects communicate and operate? Can something less rigid than the 'specification' and more open to interpretation be employed? Can we bear to engage in a process that cannot be controlled 'rationally'? Might architects begin to encourage material appropriation, experimentation and (brace yourselves) error by 'others'?

Minor planning and architecture

In her book *Architecture and the Text*, Jennifer Bloomer speaks about minor architecture[17] following from Deleuze and Guattari's theory of minor literature.[18] The broad principles

---

17 J. Bloomer, *Architecture and the Text: The (S)cripts of Joyce and Piranesi*, New Haven, CT: Yale University Press, 1993, p.36.

18 Gilles Deleuze and Félix Guattari, *Kafka: Toward a Minor Literature*, trans. Dana Polan, Minneapolis, MN: University of Minnesota Press, 1986, pp.90-1.

Open Kitchen or 'cookery architecture' 185

of this extend to the Open Kitchen. Deleuze and Guattari define the principles of minor literature:

1. '... a minority constructs within a major language. It is an appropriated language'.
2. Everything is political: '... its cramped space force[s] each individual concern to connect immediately with politics. The individual concern thus becomes all the more necessary, indispensable, magnified because a whole other story is vibrating within it.'
3. Everything takes on collective value: '... what each author says individually already constitutes a common action... [it] does not refer back to an enunciating subject (but to)... the collective assemblage of enunciation'.[19]

In this project, it is proposed that a minority group of local women, students and the university take on live projects.[20] They are amateurs, outside the profession and outside the industry. They are using a language (of building and planning) that is not their 'own', but appropriated through experimentation, construction and use. Jennifer Bloomer writes:

into this category of minor architecture, it follows, the 'ambiguous objects' fall (...) It stands in difference to major architecture, instruments of reflection of the 'divine cosmos,' Tablets of Stone with WORD OF GOD writ large (...) an architecture of desire – a minor architecture – will operate in the interstices of this architecture. Not opposed to, not separate from, but upon/within/among: barnacles, bastard constructions (*une bâtarde architecture*), tattoos (ornament, embellishment). An other writing on the body of architecture.[21]

---

19 Ibid.
20 Prue Chiles first introduced the 'live' projects into the teaching programme at Sheffield University. They are real projects, with real clients, and are intended to forge links between the university and the 'real world'. Their aim is to provide a way to operate outside of the ivory tower, broaden students' pedagogical experience and provide an education that would empower students to develop an independence from and critical view of the profession. In Open Kitchen, the scenario suggests that each stage can become a year-long live project, allowing students to understand and experiment with the realities of making a project, including procurement, resources, communication as well as design and built work.
21 J. Bloomer *Architecture and the Text*, pp. 41, 36.

Above: The Café
Right: Open kithchen scenario plans and diagrams

A *minor planning*, operating in the small gaps of the existing framework for the city planning and regeneration, operating within and between existing buildings. The nature of the process 'forces each individual to connect with politics': questioning ownership of land and facilities, questioning and determining where they are situated, what they can provide and the way they are managed. The approach to making collaborative projects between the university and local communities connects with 'politics' in its broader sense. The stories '*vibrating through it*' concern issues of authorship, relationships of power, access to knowledge, and empowerment: the know-how and means to transform a local situation.

The project becomes a political counter-point to a private development on the edge of the city; it forms a political statement regarding processes in architecture and yet is conceived as a collective authorship, it is not the will of one singular person. The methodology rests on a principle of difference and taste. Difference, existing between people, is a relationship and in this way what is constituted as a singular will also constitute a common action in a theory of difference. Without difference, the methodology no longer stands.

Through metonymy, the project provides unique criticism on current methods of planning and production of architecture. It is one that is outside the 'industry'. It involves a minor group in a minor area, producing *minor architecture*,

scenario diagram

the built products will not be timeless monuments; they will never win the Stirling Prize. This group is in a fictional narrative which operates upon, within and among existing structures of building and planning.

More personally, this project was not what was usually expected for a diploma project. It is not the beautiful piece of architectural design, so perhaps it can also be said to be 'marginal' in this sense. What was produced is not an object but a methodology and ethics for a process of architecture; a methodology based upon the celebration of difference, sharing, experimentation and collective action in the transformation of a small area of the city. It is often said that architecture is the handmaiden to the economy, and as such with their clients, architects maintain current patterns of production and ownership. Stepping away from this, the methodology suggests a means of producing architecture, where both process and product are tools themselves within the transformation of social, political and economic conditions.

It was a process of discovery. I did not know what I would produce and began the year without preconceived ideas about what constituted a 'proper' project. Through the work, I have been developing a position from which to operate in the future.[22] A lot has been cooked but nothing has been built yet. It is a small-scale, paper-based instance of a future 'what if...'.

---

22 This is significant in terms of architectural education, rather than having a personal significance alone. Jeremy Till writes: *There is a familiar complaint from the architectural profession about architectural education. 'You are not preparing the students for practice.' To which I reply, 'Which practice?' Underlying the question is the assumption that there is a single model of practice to which the profession aspires, and it is the task of education to supply students who will passively serve and support this model.* (J. Till, 'Five Questions for Architectural Education', paper presented to the RIBA, May 1997)

Chapter 12

Ruth Morrow

BUILDING CLOUDS
DRIFTING WALLS

Sherry Ahrentzen; Paul-Alan Johnson; Karen A. Franck; Leslie Kanes Weisman

Chapter 12   Ruth Morrow

JDS,
LS

'BUILDING CLOUDS DRIFTING WALLS' was a series of pedagogical events which took place over three years (2000-3). The events were part of a strategic rethinking of a first year design studio in architecture.

Like much that is social, the surface outcome is familiar and like much that is radical, the ambition and complexity of the course sometimes weakened individual outcomes. So, in order to be seen and valued as a deliberate attempt to challenge traditional pedagogies, the 'events' had to be documented and presented as they were intended; that is, as a singular assault on the conventions of early-stage design teaching.

The final production of these events took the form of a pink booklet, the realisation of which mirrored the ethos embedded in the design studio.

The university in which this pedagogy grew is one of the UK's red brick institutions: conservative, hierarchical and research-led. Within such universities there are still some efforts to encourage and support teaching. One mechanism employed is to give awards for teaching excellence. Excellence is measured by the number and skilled usage of the pedagogical phrases used within the application form: this is somehow understood to be a more effective method of evaluation than asking the students. Typically the award is given to an individual, more precisely one full-time academic, a reference perhaps to the more classical days of teaching and the lone voice behind the lectern. For their diligence, the awardee is given a sum of money to spend on a teaching-related activity on which a final report must be submitted. Presumably, this is intended to provide further evidence of their teaching excellence, but in reality it simply adds to the burden of an under-resourced, over-stretched and poorly valued individual. In this instance, there was a certain irony in an abstract and individual-centred system rewarding such a collective and responsive approach to teaching.

But feminism shows us that progress can happen despite the system and that most of these gains occur through intelligent subversion. So, as I accepted the teaching award from the red brick university, my part-time colleagues in the School of Architecture reminded me that it was our combined efforts that made the design studio what it was. My job, as the full-time academic, was both to clear the way (deal with the bureaucracy of teaching) and cover the tracks (deal with the bureaucracy of teaching) and theirs was to bring enthusiasm and sufficient doubt about architectural education to make each step both critical and considered.

So the teaching excellence award financed the production of the 'pink booklet', paid some diploma students to design the layout, and more importantly, paid part-time teaching staff for their contribution of text and time.

While our teaching was in part the result of external forces (reducing resources, increasing student numbers, and additional layers of bureaucracy), a very distinctive and creative *mood* in the school of architecture fuelled and focused our efforts. The school was and still is populated by staff who want to introduce more critical concepts of architecture to the student body. In finding ways to formulate this for students, we absorbed it naturally into our own teaching practices. Perhaps the most relevant example was a lecture that grew out of a collaborative effort of three academic staff. From various readings, we summarised the characteristics of a 'feminised space' and 'feminised practice':

Feminised Space

Process as well as product

User as well as used

Experience as well as Knowledge

Content as well as style

Temporal as well as Frozen

Everyday as well as iconic

Multi-sensorial as well as visual

All-the-spaces-in-between as well as the public/private divide

Feminised Practice

Collective, not always individualistic

Listening, not always telling

Facilitating, not always controlling

Non-hierarchical, not always Hierarchical

Fluid, not always fixed

Doubtful, not always Certain [1]

1   See, for example: Sherry Ahrentzen, 'The F Word in Architec-ture: Feminist Analysis in/of/for Architecture', in Thomas Dutton and Lain Hurst Mann (eds), *Recon-structing Architecture: Critical Discourses and Social Practices*, Minneapolis, MN: University of Minnesota Press, 1996; Paul-Alan Johnson, *The Theory of Architecture: Concepts, Themes & Practices*, New York: Van Nostrand Reinhold, 1994 (notably Chapter 4, 'Professional Attitudes', and Chapter 5, 'Feminism: Shifting the Agenda'); Karen A. Franck, 'A Feminist Approach to Architecture: Acknowledging Women's Ways of Knowing', and Leslie Kanes Weisman, 'A Feminist Experiment: Learning from WSPA, Then and Now', in Ellen Perry Berkeley (ed.), *Architecture: A Place for Women*, ed. Ellen Perry Berkeley and Matilda McQuaid (Washington and London: Smithsonian Institution Press, 1989)

These characteristics also offer the clearest summary of the teaching content and methodologies described in the pink booklet, activities that were framed by a feminist pedagogy of inclusion.

The following text illustrates how some of these characteristics manifested themselves in our learning and teaching activities.

Process as well as product
Experience as well as Knowledge
Listening, not always telling

The first project, first week of the first year design studio did everything that it should never do. It wasn't an abstract project, narrowly focused or skill-based. It didn't challenge students to rethink their perceptions of architecture, instead it kept the floor firmly beneath their feet. The first project allowed the students to do exactly what they thought an architecture course would ask them to do. In one week, they found a site and made a proposal for a building for themselves. There was no brief and no restrictions, other than they had to work in groups and build a model. Their resulting designs, which could almost singularly be described as 'sprawling wish lists' of swimming pools, bars, saunas, etc., demonstrated why such a project rarely happens in a school of architecture. Believing architectural design to evolve from plans, the students left the model building too late and presented rather scrappy first attempts. Not exactly the sort of work one would show a visiting critic!

Nevertheless, it was an exciting project for students because their first experience was taking part in a project that matched their perception of an architecture course. The pedagogical aims of the project were, however, more complex. We used the project to watch and listen to the students – getting a measure of them before we started to 'tell' them anything; it was also an effective way for students to get to know each other since negotiating over space is one of the most primal activities humans can be involved in. And of course, the product is not the building but the opportunity that the

Building clouds, drifting walls

product gave us to talk about the process ahead, i.e. the process of learning to become an architect.

PROJECT ONE: HOME TRUTHS
DESIGN A BUILDING TO SUIT YOU AND YOUR FELLOW STUDENTS' NEEDS. GROUP WORK 1 WEEK
EXPOSING THE CURRICULUM.

90% of the time, at least 80% of the answer lies in your head.

STARTING WITH WHAT YOU KNOW

Architecture is not a subject taught in UK primary or secondary schools and yet we all use and abuse the built environment every day. So should first year architecture students consider themselves novices or experts?
The problem with adopting an "empty vessel" approach is that by denying or misusing your own experience a vacuum is created that may become filled with other people's assumptions, regardless of their value.
Focusing on domestic space during first year allows students to use their knowledge of a type of space most intensely experienced by themselves firsthand. Our homes / rooms are one of the few spaces that we are all able to influence directly and manipulate for our own purposes. When asking students to design a place to live in Sheffield for themselves and their design studio group, both the location and 'family group' are familiar: privacy, sharing, individuality, eating, sleeping, relaxing etc. Successful group working is encouraged so that individual students bring as much of what they know to the project as possible.

WE DON'T KNOW, WHAT DO YOU THINK?

Most of the students that begin in the first year are used to a model of learning (found in school) in which the teachers are in charge, and are seen to hold the knowledge that the students require. Learning is a matter of passing on this information. In contrast, first year aims to enable students to take responsibility for their own learning and develop their own understanding of architecture and its production. This new position can provide discomfort for new students who are looking for 'answers', but fundamentally acknowledges the subjectivity of architectural understanding and the individuality of approaches to it.

As tutors, we acknowledge that we have more experience in design and we hope that we can use that experience to help students to find their own way, but ultimately we are also hoping to learn from them. This is why we actively promote and facilitate workshops, projects, discussion groups and project reviews run by students. The introduction of clients / certain projects also means that students come face to face with another set of perspectives and opinions, perhaps in contrast to tutors' views on the project, serving to remind students of the subjectivity and individuality of their tutor's views.

PROJECT TWO: DESIGNERS, KINGS AND ASSEMBLERS
DESIGN AN OBJECT USING CARDBOARD, PRODUCE CONSTRUCTIONAL DRAWINGS. PASS ONTO ANOTHER STUDENT TO BUILD DESIGN. 2 WEEKS

The project 'Designers, Kings and Assemblers', occurred in the second and third week of the first semester. In the first week the students were asked to design an object that they needed in their lives. The object was to be built of a sheet of standard sized cardboard, of which a sample was provided at the start of the project. By the end of the first week they were asked to produce a set of drawings, from which someone else could construct their design.

'...you knew what size, what material and how it was constructed. It was quite realistic - someone's actually going to have to put this together.'

'Only being able to communicate your ideas through a drawing was really difficult. It really made you think.'

In the second week the students moved from being designers to assemblers. They were given the drawings from another student in the year and were asked to construct the design.

'Enjoyed the task of building to a design - necessity to be precise and accurate - but also found it restrictive for the same reasons'

'...found it hard to achieve same enthusiasm for building someone's model as I would have had, had I been building my own'

The assemblers were told to 'mark-up' the drawings with any altered or missing information and were allowed to 'call the designer to site' to discuss changes. Once the object was completed the designer was asked to complete a questionnaire, recording the assembler's opinions of the drawings, method of assembly and design.
The project ended with small group discussions on the architect's responsibility to integrate the processes of design, representation and construction. There was no need for a formal 'crit'.

The project review was used not to review the work in itself, since the students could observe and compare the quality of their work with others, but rather to discuss 'What now do you need to learn?'. They built a curriculum and we listened. Of course, their curriculum matched very closely the one that was already in place (it had to be in place – that's the way the modern university works) so we didn't discuss the overlaps – why talk about things they already know? Instead we focused on the areas that were either fewer well described or completely missing. Two missing areas always stand out in this process – the first is learning about other people's needs. In respect to this, students typically only list the need to learn about disabled legislation. The other gap in their perception of architectural education is 'learning to design'. Over the three years that we ran this project at Sheffield, fewer than 10 students out of over 300 listed that they needed to learn to design. This allowed us to quickly expose one of the design studio's most persistent and pervasive myths, i.e. that design ability is a 'gift' rather than a 'skill'.

At the end of the process, they have made an extensive curriculum. We encourage them to see that collectively they already hold a wealth of information. While studying architecture may be a new experience for them, that their direct experience of the built environment over the past 18 or so years (and in the years to come) is as valuable as the knowledge they will acquire within the academy. We ask them to keep alive the memory of being an *ordinary user*.

Everyday as well as iconic
Non-hierarchical, not always Hierarchical
Fluid, not always fixed
Facilitating, not always controlling

The pink booklet also contains a description of another short pedagogical event entitled *Designers, Kings and Assemblers*, which ran over a two-week period. In the first week, the students adopted the role of the Designer, designing an everyday object they need in their lives. This brief is reflective of the belief that engagement in the project is heightened when the brief fulfils a personal need. They were asked to represent the object two-dimensionally, in such a way that someone else could build it. The only limits were that the object must be constructed from cardboard. In the first week, the project is a traditional and an obvious beginning to a first year design studio project.

In the second week, the student takes on the role of assembler and, using the 2-D representations of another student, builds that design. It is a project that replicates on-site frustrations and week 2 can result in heated discourse. At the end of the two weeks, the review is addressed, only in part, to a discussion of the end product and tutors are able to draw the students' attention to such things as: the frequent relationship between 'tight brief' and 'potent outcome'. More time, however, is given to reflection on the interdependence of design and representation, designer and assembler. Students recount their experience of being the assembler and are asked to consider what determined a successful form of representation. Issues of clarity and accuracy are obviously covered but value is also given to representations that engage the assembler in other ways, for example, in terms of attractive-

Building clouds, drifting walls 197

ness, informal styles of communication, or presentations that set a framework and criteria allowing the assembler to make decisions that draw on their own expertise and build a sense of ownership. The discussion reveals that skills, effectiveness and choice of appropriate methods or orders of communication contribute to the overall effectiveness of the design.

But the process also reveals to students that, in certain projects, the role of 'king' can change hands between assembler and designer. That an assumed hierarchy (of architect telling tradesman) is not always useful and that relationships can be equally productive when there is a certain amount of fluidity.

Of course, there are examples of students who work long hours to produce highly accurate, step-by-step drawings that allow neither for alteration nor error on behalf of the assembler. This approach frequently works and the final product is just as the designer imagined it to be. But, we then discuss the cost of this approach: long hours, low wages, and an overly controlled process that negates the expertise and creative potential of the other.

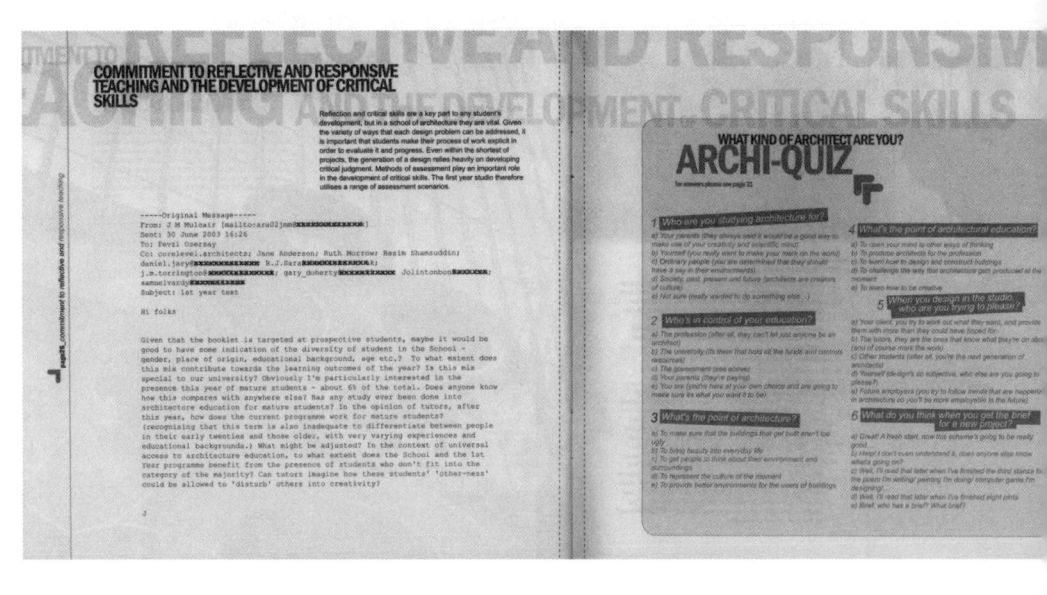

Temporal as well as Frozen
All-the-spaces-in-between as well as the public/private divide
Collective, not always individualistic
Doubtful, not always Certain

The event *Four Days on the Outside* (full name: *Four Days on the Outside and One Day Back in the Academy*) was designed to expose students to external conditions. Tutors were identified, mostly external but some groups were also memorably led by mature students. The brief was that each group should either work *with* or in ways *visible* to people 'on the outside'.

Four days is not sufficient time for works of great architectural gravitas to emerge but short projects demand fast responses and high levels of creativity. In the early stages of design education, when students are still intellectually uncluttered and naïvely motivated, they are well able to match this challenge. So these four days are intensive and surprisingly productive.

Pedagogically, this event also encouraged students to 'take action'. It dared them to intervene in the public realm and in so doing their actions revealed the politics of space. As a group, they had to negotiate sustained access in order to carry out the work. Often they realised that access is sometimes easiest in the unclaimed spaces of the city – the spaces that fall between public and private – and that their temporal actions can be as powerful as fixed or frozen projects.

The resulting projects varied enormously. One group designed a pyramidal structure from recycled tyres that acted as a drink-can disposal structure in the city centre. The project's revelation was who owns and 'surveys' public space. Another group worked with a wildlife trust and local housing residents group to devise phased strategies to regenerate a piece of wasteland. One group, led by diploma students, occupied a courtyard shared by a range of craftsmen. They devised a project to clear up and recycle rubbish in the courtyard and in so doing created a system of bartering materials that revealed the inherent value of materials. Other groups made sketch proposals for specific user groups, occupied spaces and/or built structures that draw public attention.

Alongside all of the projects, one group was formed to specifically document the entire event. They played a significant role on the fifth day, i.e. the final day back in the Academy, where they present a digital overview (often in the website form), and a commentary on the four days. The floor is then opened to students, external people and others to comment on the relevance of these actions to the practice of architecture.

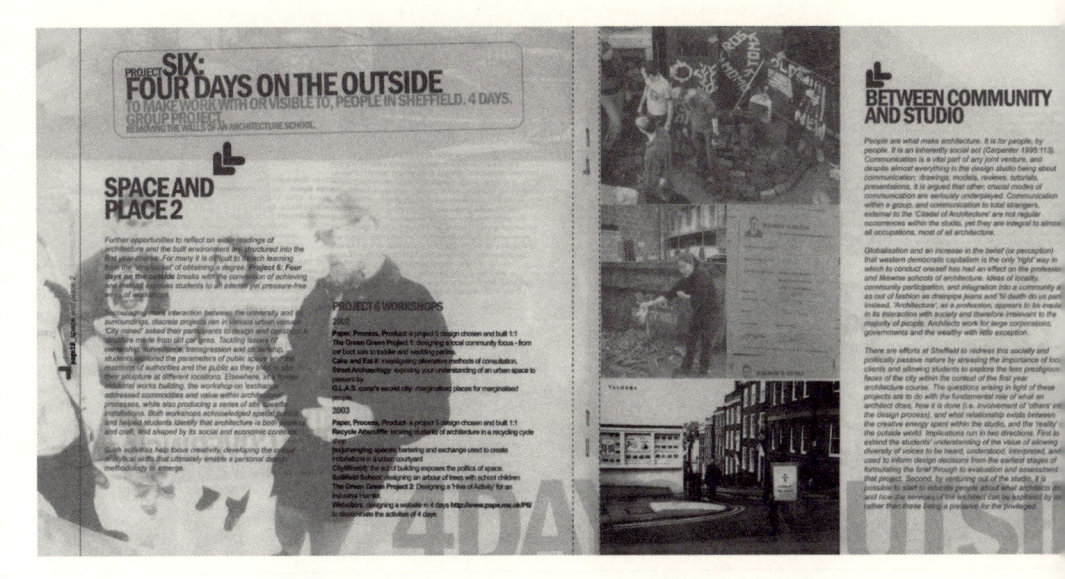

Such actions meet the wider aim of the event, which is to raise the profile of architecture in general and the course specifically. We wanted to ensure that people beyond the course knew of its existence and indeed its value as a conceptual machine. The project aimed to develop relationships that offered future support; creating an event where learning and action in the public realm became a shared activity between studio and community, between university and city.

In conclusion

This selection of pedagogical events all share similar characteristics. They have a 'surface' pedagogy that resembles traditional teaching projects in schools of architecture. Underlying each event, however, there was a pedagogical urge to provide students with not just knowledge and skills but also an experience that questioned their own and society's values. These events sought to expose rather than impose values.

To do this, a space-time for discussion was very deliberately set aside (the 'fifth day'). This was more than a period of reflection since we looked forward as well as backward. It was a time of exploration, examining what the outcomes of the events were, what products other than 'designed things' resulted from the process and suggesting ways that this informed the position of architecture, architecture courses, and architects to society.

In doing so, the complexity of these positions was exposed and of course the varying stances and in many cases outright contradictions that exist in our structures. From such discussions, there emerged an alterative way to re-conceptualise architectural practice, and indeed architectural education, as a process of managing conflict rather than necessarily trying always to resolve it.

One crucial step in managing conflict is not to hide it away. Discussion of process and issues requires explicit documentation and it was the development and repetition of this skill that naturally led to a *pink booklet*.[2]

---

2   Next step... The author is currently course director for a new undergraduate course of architecture at the University of Ulster in Belfast. When establishing a new course, one question remains at the fore: where does architecture begin and end? Images X and Y show one event from the first semester of the first year of the course where shadows on canvas offer one end of that spectrum. These images are curiously illustrative of the motives that drive the new course. Thoughts about the interaction of people with the material, regardless of the conflicts, contradictions, messiness and unpredictability of those interactions. The new course hopes to make room for an un-sanitized version of 'everydayness', seeking opportunities to place the internal dialogues of architecture, quite literally on the outside of the 'school'. Doing, building, acting upon ideas – 'putting it out there' allows us to observe how, when internal dialogues are exposed, the reactions of others enrich the meaning and progression of the internal dialogues.

This is creativity not under control in the 'designed' sense, but rather, one that springs from examining and managing conflict. It is a relevant and timely approach for a new architecture course that is based in a city emerging from troubled times, when being politicised and creative was quite literally 'dangerous'. Within this historical context, the new course seeks to make being creative, active and *in open dialogue* a mainstream activity for all its participants: students, staff, observers, helpers and, as such, we would welcome comments and ideas from all who support progressive and inclusive change in architecture education. r.morrow@ulster.ac.uk

Thanks to all the feminists (m +f) who have inspired our teaching.

Chapter 13

Ilana Salama Ortar

URBAN TRACE
civic performance art and memory in public s

Chapter 13　　　　　　　　　　　　　　　　　　　　　　　　　　　　　　　　　Ilana Salama Ortar

URBAN TRACES: civic performance art and memory in public space

Civic Art is an engagement, in the form of an artistic intervention, with individual cases of social deprivation and political constraints. It involves reflection on notions of public space, public opinion and their relationship to the *tabula rasa* and the present considered from the space-time point of view, in which the individual's inherited context and collective memory have disappeared.

In order that the people engaged – Israeli, Palestinian, immigrant or displaced persons – do not become irrevocably estranged from the other, alienated from themselves and the world, we must evoke and develop memory.

Civic Art gives a voice and tangibility to the processes that efface memory in public space, so that memory itself can be a practical, mobile, playable, transformable accomplishment.

In order to get the 'archaeological dig' underway and to develop political/poetic research in these places of memory, I make interventions in their spaces. This involves developing channels of information, interaction and questioning as well as parody models or critiques – all clues whose aim is to open up and reawaken listening and speaking.

In my Civic Performance Art projects, I aim to explore the effectiveness of performance art in taking a significant role in the public discourse on political and socio-economic issues, opening up another channel through which people's voices can count. The main innovation I propose has to do with the citizens' own role in the performance. They are perceived as actors in a play, without whom there can be no performance. For me, the participation of such citizens is part of the performance itself. This is a step towards a partnership characterised by mutual responsibility.

A project

The Tower of the Prophets and the Villa K'houryi, Wadi Nisnas, Haifa, Israel – the Visible and the Invisible in Israeli and Palestinian memory: five installations of Civic Performance Art 1995-2000 in the shopping centre and vicinity of Wadi Nisnas, the old city which encompasses the centre: street, Protestant church, the Garden of Memory and the Jewish and Arab community centre.

These civic art installations are intended to make tangible and audible the processes of suppression of Palestinian memory in urban space. It was through the history of a modern building in the city of Haifa, a public building including offices and a shopping centre, that I decided to speak of the conflict between two communities of different cultures, living in the same urban locality, two communities who must find a way to communicate, despite everything, against all the odds. Finding a way of living together is the crucial challenge of a complex, ambiguous and often contradictory urban reality. My way to assist my fellow citizens was to use a montage of forms to show the range of attitudes that are taken every day in the face of conflict.

It was a public objective work, open to interpretation and discussion.

The social life of Haifa has been dominated, at least since 1948, by the serious Jewish/Arab division, just as international relations in the Middle East have been dominated in the same period by the serious, indeed bloody, opposition between the State of Israel and the Arab States.

The division is generally understood as a religious conflict: there are Jews and there are Muslims. But this polarisation covers up a much greater diversity. We can all too easily forget that among the 'Arabs', there are also Christians and that even among the Christians there are orthodox Catholics, Maronites, Protestants and others. Similarly, we forget that among Jews there are Orthodox and Reformed, Sephardic and Ashkenazi Jews, as well as atheist Jews. These distinctions are numerous and are not all of a religious nature. They always lead to a past, whether recent or further back, and call on stories, mythologies and geographies that are sometimes near, sometimes far away. But they share one thing in common: all come under the strict Jewish/Arab opposition.

Haifa, like Jerusalem, is a city with a great religious tradition. It is even known as 'the city of religions'. But, unlike Jerusalem, it does not hold an important role in the geopolitical imagination of the modern world. It is not under the gaze of the media spotlight. This is a piece of luck. Within Haifa, it is still possible to feel differences co-existing. It is this value of co-existence in peace that I was hoping to unlock and affirm with an artistic project. The differences between the ethnic and religious communities are historic. But Haifa is also a modern city. It is a business city on a local, regional and world scale. In the utopian vision of the future, the unifying activity of commerce would replace ethnic and religious distinctions.

In fact, we know that the world economy is attempting to make cultural life more homogenous. But in a city like Haifa in the North of Israel, at the crossroads of Lebanon, Syria and Jordan, with a very mixed population, can we believe that forgetting differences will ever be anything other than 'repression'?

As the carrier of great hope, globalised commerce also represents the threat of a levelling-out, a blindness, a race headlong into a future of renewed conflict. It was for this reason that I specifically decided to make an intervention in a shopping centre. A shopping centre is a shared daily expression of modernity and its global extension. Here all identities, all historic experiences, can co-exist under the same roof, even though they are forgotten, suppressed and hidden by the screen of merchandise. It is a site where the destiny of the city and by extension the destiny of the entire region of the Middle East can be tipped out of balance. It is a prototype site for all societies who live with internal conflict: conflict of religion, ethnic conflict, or land conflict.

The building in question was built in 1979, after the local area had undergone a major restructuring process that might be termed an attempt to make a 'clean slate'. The building is a modern one, monumental in scale – a geometric volume clad in mirror glass. Its conception goes back to the end of the 1960s, when the city council decreed a programme of urban restructuring. This building was to be iconic of a city and State that were recognised on the world economic stage. However, the construction was delayed by ten years or so and this gave it a slightly ghostly look, even in spite of the violence of its siting in a historic context. The local residents are part of a marginal population: Jewish and Palestinian working class, a minority in the country. Behind the shining architecture of the modern tower, we find that the site played a crucial strategic role in the war of 1948. The fortress which preceded the current tower was called Villa K'houry and was the last stronghold of Arab resistance in the city. Its fall sealed the victory of the Jews of Haifa and was the start of the flight of Arab civilians to surrounding countries. So the site of the present-day tower is actually a symbolic landmark, evoking for Palestinians the feeling that they are victims of the conflict and for Jews a feeling of guilt, especially among those of my generation who did not live through the war but who have been brought up on its myths.

We might even say that it is the building itself that reveals or hides the complex urban fabric of the site. This building, like a vast optic machine, can reflect all the facets of the conflict dividing Israel, both on the exterior and in the interior. To clarify the conflict for those caught up in it, it is important to uncover repressed history in 3D forms treated equally. The best way I found to make history visible and to recover memory was to go back into the shopping centre and use the built space and a theatre of the past that rises up in the present. That is to say, arrive at a situation where the building itself tells the story of the city.

The history of the war has left an aggressive, violent and tense legacy, still found in the area around the tower. It is a dying area from the urban and demographic point of view. The alienating presence of the modern tower on such an ancient site has created an aggressive space, a violent rupture in the urban fabric, a mixed population which cannot find ways to speak to each other. But to our great surprise, as soon as you go into the shopping centre, a human balance is re-established. Activities and expressions no longer allowed on the street happen once more inside, in the maze of streets/corridors/pavements that link the boutiques and shops. Here human relationships are transformed in a fast rhythm. The local population finds the beginnings of a common language in the neutral space of a commercial environment, a 'non-space' or 'atopy', to use Foucauldian terminology. The shopping centre functions on the basis of a negation of history but it is also here that the living diversity of history is seen most clearly. Therefore, it was here, in this non-place, that I decided to 'sell my wares', to bring together precise factors of repressed history by means of a model, cards, photos, videos. 'To sell' here means laying out in a strictly equal way all the historic points of view, all the layers underlying the present situation (which is, in fact, a sales situation: a shopping centre).

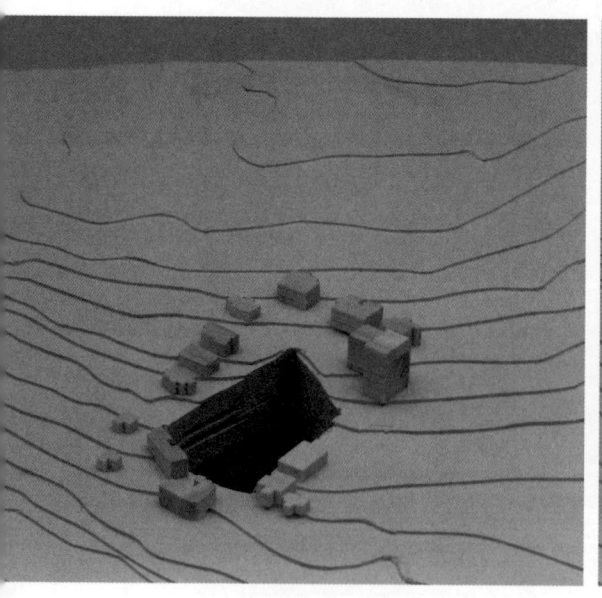

The main item: a double model representing the Arab fortress encased in the transparent structure of the modernist tower. The proportions of the two elements reproduced exactly their relationship of scale as well as their geographic setting and their construction materials.

What is an artistic model? It is not a real situation. It is a reduction, a miniature, that can be manipulated, that becomes the backdrop for a mental, imaginary experience. It is an opportunity to look at reality in another way, to speak with another about this different looking. In this first exhibition my experience consisted of evoking the past in the present. Villa K'houry was brought back into the heart of the shopping centre. The monument of the repressed past was put into the modern monument, in order to suggest how modernity can accept the ambiguities of its history. To be in the present means to also recognise the past. I wanted to see the Tower of the Prophets as a monument to the present, a present that is rich with its past. And when I exhibited the double model, with the villa and the tower, just like an item for sale, I hoped to see the exchanges, even commercial ones, as exchanges between different human beings.

This double model was sold as a 'package deal'. It was shown on a stand in an empty shop, exactly like a 33cm television or a scaled-down sports car. Selling this double model led to at least three readings by the potential consumer, the one about to make a personal choice:

- Buying it, they are investing in the plural history of the city, even of the territory that is today called Israel; they are writing themselves into the complexity of what exists.
- Buying it, they fall in love with just one of the two elements and exalt it as a symbol of their oppressed or victorious people; it affirms an aspect of identity.
- Buying it, they acquire a banal consumer item that will be thrown away once it no longer provides amusement; they choose to cancel history.

The environment of the modern tower, which is neutral in regard to historic conflicts, gives the individual the freedom to choose. The miniaturisation of the building and its temporary presentation in a commercial context are a metaphor for openness, for the possibility of physical and mental change. But it is clearly about an ethical choice, focused on society, not on simple individual gratification.

Other items, documentary souvenirs, were 'sold' in the shops in the centre. On the one hand, there were models of the model, 'history keys' that developed a type of tourist logic of memory-merchandise, fetishised history; on the other, documentary photos and cards as collages were framed according to a system. The uniformity of presentation in all the shops was to highlight the fact that the subject of the exhibition was the space of the whole centre.

In the labyrinth of consumption, the framed collages marked out a path, an itinerary, transforming movement from one shop to another into a steady journey through history. Memory must be sought out, it must be reconstructed, since it gets lost. Crossing the threshold of the shopping centre, the consumer received a map indicating a journey that would show the evolution of the site and the city through consumption, or vice versa.

Here again, the act of buying is linked to history and the social context.

Difference in Haifa is first seen in groups, either ethnic or religious. These groups can be separated by identities, points of dispute, questions of rights and resources to be shared. Nothing replaces negotiation for solving such questions. But in order to negotiate, one must respect the other. And in Haifa, this also means respecting the other as a member of a group. Community identity is often a closed sanctuary, 'protected' by aggression. But it can also offer the ground for a generous citizenship. If the exhibition in the Tower of the Prophets went well, it is due to the fact the local children took huge pleasure in finding their house, their place of origin on a map of Wadi Nisnas, an aerial photo exhibited on the bare ground. The pleasure of the children attracted the adults who could see the shopping centre as their place, filled with their living history, but also shared, crossed by the human diversity of the modern city.

This complex experience of identity and otherness in a single place – in a single city, in a single world – is at the heart of daily life.

The goal of an artistic project is to make this experience tangible, conscious and to give it value as an alternative to conflict.

The next step was to bring the outcome of this historic sharing to different places of worship around Wadi Nisnas, to deepen the experience, to enrich Haifa's multiple pasts and to contribute to the recognition of their traditions as an integral part of the present. To this end, I started a series of small exhibitions which take up elements of the first, each time working through the specificity of the site and its actors. I wanted to have Jews and Arabs, but also Sunnis, Orthodoxes, Sephardics, Protestants, Catholics, Maronites and still others. All have a part in the city and they can be complementary one to another.

It is this complementarity that I look for when I combine modern architecture with sanctuaries/sites of community identity. It is a real research process, involving an ever-growing number of partners. In this process the results of the research are provided by debate and exchange; instead of a simple historical or sociological report, we aim to build up a narrative of the city of Haifa. This narrative, which was begun in the shopping centre, is based on multiple movements.

When the series of exhibitions-exchanges was finished, the following exhibitions were held:

- *Otherness* – the English hospital, alternative venue of Villa K'houry (December 1995)
- *Hassan Shukri Street* – The Centre as a Chink (December 1996)
- *Memorial Park* – a 'tableau vivant' aimed at staging the subject of individual memory and collective memory, based on rememberings of the city of Haifa on the national day of independence (May 1997).

>The results of this series of exhibitions-exchanges – personal statements, suggestions and questions, collected in many forms (texts, photographs, objects, videos, database) – were presented in a small modern construction, a kiosk – a temporary pavilion in front of the city hall and in the shopping centre. These constructions were clearly a monument to the present, that is, the recalled multiple pasts which flourish in the present and which go beyond any one identity to make a city.

>All images in this chapter are artwork of the author.
>Translated from French by Jacqueline Mulcair

Chapter 14

Pauline Boudry
Susan Sontag   **Marion von Osten**

Ursula Rabe-Kleberg

Martine Anderfuhren

SEX & SPAC
space / gender / economy

David Harvey
Christine Boyer

Jane Jacobs

Mary McLeod

Chapter 14   Marion von Osten

H. Hitz, R. Keil, U. Lehrer, K. Ronneberger, C. Schmid and R. Wolff; Cynthia Cockburn and Ruza Fürst-Dilic

Ulla Kilchenmann

Roland Barthes
Walter Benjamin
Frigga Haugg

E:

Yvonne Doderer

Katja Reichhard

Konrad Wünsche; Martine Anderfuhren

Sander Gilman

Susanne Bachmann

SEX & SPACE: space / gender / economy

Introduction

The project *Sex & Space*[1] analysed the architectural, urban and planning realities as representational systems of gender-specific attributions and looked at them in terms of their potential for change through cultural practices. In conjunction with architects, artists, activists and theorists, we investigated commonly held ideas about the public and the private realms and their traditional interpretation as 'male' and 'female' respectively. In *Sex & Space*, gender difference was not seen as innate and biologically determined by binary oppositions. Instead, our intention was to investigate how space as a social, cultural and political structure produces specifically gendered identities and how the construction of difference has been inscribed in our notion of space. Polarisations in social space and segregation on ethnic and gender-specific grounds were not only seen as an effect of macro politics, but also as part of a specific discourse that causes not only gender difference but also Eurocentric subject positions. One of the project's main questions was: what kind of theoretical and artistic tools might be developed to deconstruct these opposites?

When starting the project in the Shedhalle Zurich in 1996, we – Rachel Mader as art historian and myself as curator – initiated a reading group in order to discuss this issue with artists and architects, allowing us to look into the problem from varying points of view. The reading group came together

---

1   *Sex & Space* was an event with an exhibition, a production studio, lectures and films held at Shedhalle Zürich, 6 September – 6 October 1996 and at Forum Stadtpark Graz 10-26 October 1997.

Sex & Space I + II was curated by Marion von Osten. Participants included: Fenja Abraham (CH), A-Clips (D), Martine Anderfuhren (CH), Students of the Faculty of Architecture, TU Graz (A), Hatice Ayten (A), Susanne Bachmann (CH), Simone Batschelet (CH), Jochen Becker (D), Bettina Behr (A), Serap Berrakarasu & Gisela Tuchtenhagen (D), Pauline Boudry (CH), Commission for Womansrights, Graz (A), Yvonne Doderer and Cora Schafer (D), Ines Doujak and Gabriele Marth (A), Lukas Duwenhägger (D), Dominick Eichler (D), Julian Goethe (D), Frauenhaus Graz (A), Frauen Lobby Städtebau (CH), Frischmacherinnen (D), Edith Glanzer (A), Irmi Hanak (A), Andreas Hofer (CH), Innenstadt AG Zurich (CH), Ina Ivanceanu (A), Lea Jaecklin (CH), Sylvia Kafehsy (A/CH), Anke Kempkes (D), Dorothea Kress (D), Brigitta Kuster (CH), Pia Lanzinger (D), Elisabeth Lechner (A), Susanne Lummerding (A), Rachel Mader (CH), Mascha Madärin (CH), Ariane Muller (A), Irene Nierhaus (A), Marion von Osten (D), plattform & Ulrike Kremeier (D), Katja Reichard (D), Marie Ringler & Meike Schmidt-Gleim (A), Stefan Römer (D), Susanne Sauter (CH), Pia Siegrist (CH), Cornelia Schmidt-Bleek (D), Jo Schmeisser (A), Anni Staehlin (CH), Nicol Stolz (CH), Josef Strau (D), Gertrud Strempfl (A), Vor der Information (A), Irene Windisch (A), kunstverein W.A.S. (A), Alexandra Wurz (A), Ina Wudtke (D), Sandra Zech (A), Barbara Zibell (D), Michael Zinganel (A).

More information on the project can be found in the two Sex & Space fanzines, which can be ordered at the Shedhalle Zurich or Forum Stadtpark Graz, or at *www.k3000/sex&space*

Sex & Space I, poster, 1996

six months before the start of the exhibition and was to have a strong influence on it. Apart from panels, lectures, workshops, city tours and interventions, the preferred media of artistic and discursive output turned out to be film, photography and text. For four weeks, the exhibition space served as a production studio for counter-representations. We imitated real TV and movie sets, thus highlighting the matter in question. The exhibition hall acquired the character of a constructed and artificial space. The fake production studio was set up to realise non-hierarchical productions and encourage collaborations, and tried to make visible forms of everyday work that usually go unnoticed.[2]

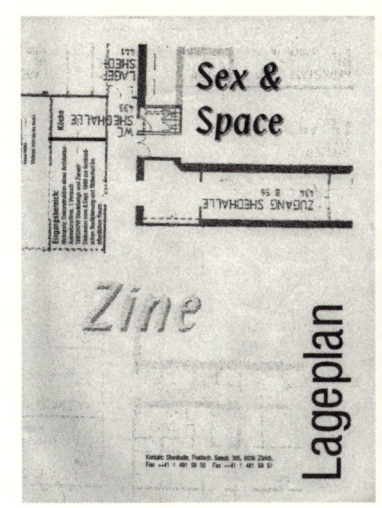

Fanzines; *Sex & Space I*, *Sex & Space II*

In the second part of the project, which was organised in collaboration with the artists-cum-curators Pauline Boudry and Michael Zinganel in the Forum Stadtpark Graz/steirischer herbst '97, the aim was not simply to adopt the exhibition format, but to continue the work on site with committed people. The topics explored in *Sex & Space II* ranged from spatial determination of inequalities of gender, race and class in current and historical politics, planning utopias and their social techniques to the question of how working spaces and work itself represent and determine the social status of human beings. *Sex & Space II* presented these issues within a new exhibition section that referenced city planners' exhibitions in an ironic way. Apart from organising theoretical and practical workshops, lectures, films and interventions in public space, *Sex & Space II* collaborated with local feminist groups and organisations. During the two projects, we tried to find out what devices were possible for an artistic and theoretical breakdown of anti-urban, sexist and racist standpoints. We

2   The project *Sex & Space* referred to a notion of cultural practice where art and the social and economic relationships created in an art institution are seen as part of social reality. This is a fact that more often than not is left out of consideration, since art is mostly viewed as an isolated aesthetic product. Since 1994, the Shedhalle Zurich had been involved in projects that implied the confrontation between political conditions and the given factors of art production and perception, mostly from a feminist point of view. Artists and cultural professionals collaborated in the developing of most of the project concepts, thus breaking up the prevailing role models and hierarchical labour divisions of curator/artist/theorist.

sought to reassess the concept of the problematic infringement of public space by violence and sexual harassment in conjunction with people from the fields of architecture, art history and politics, breaking with the stereotypical role models, deconstructing the construction of 'difference', and defining the city as an artificial place where identity roles are not fixed and prescribed.

'Paris is a woman in the prime of life... But Berlin is a girl in a pullover...'
'This is the new starting point, we have to begin from here...' – comes a whisper from off-stage, commenting on the final sequence of Jean Luc Godard's film *Deux ou trois choses que je sais d'elle*. We are looking at the model of a town placed in a meadow, assembled from packaged consumer goods: food, detergent, coffee, cosmetics. The message is clear: the town as such is a commodity and social realities are conditioned by capitalist mechanics of exploitation. Only by accepting this, will we be able to change our habitat called 'city', and hence the social conditions, which are not only caused by the city, but also have an effect thereon. The city is a commodity, a consumer item, a product of the capitalist system of the West, as are its inhabitants. They are an integral part of this

Städtequiz in *Sex & Space I*

exploitation machinery, and are therefore faced with the constant need to prostitute or sell themselves, whether they are film directors, housewives, workers in the factory or employed in the service sector. They sell themselves, their bodies, or their work time, in order to be able to consume again – so the film seems to claim. The final sequence, the model of a city made up of consumer goods, constitutes an invitation to act: we, the spectators, are meant to bring about a change in those conditions; the means, however, remain within the domain of political fiction.

This film, produced thirty years ago, summarises some of the central assumptions of European left-wing city sociology. The film claims that urbanity has a formative influence on the subject, that urban development projects (involving architecture and planning) have to be interpreted as structural for the existing conditions of power, and it demonstrates what roles urban capitalist space allocates to us.

As early as 1973, the geographer and city sociologist David Harvey noticed, in his book *Social Justice and the City*,[3] that if space and society are looked at as two separate categories, the assumed preconditions are wrong. Harvey, and later other city sociologists, saw spatial organisation and division as a product of social actions, conventions, structures of thought, and power. Furthermore, he stated that the social status of a person is a factor in the consideration of spatial organisation, urban structures, architecture, and the concept of public life and privacy. At the same time it is the place where social values are lived and reflected.

But urban space is also defined and characterised by popular representations, fictions and narratives. Metropolises in particular have been part of a complex production of myths. Depending on the social context, cities have been given various names: megalopolis, metropolis, Moloch or 'jungle'. These fictions and narratives describe the city as an imaginary object, marking its character through discourse and ideology. In this way, an image or a representation of a city can be maintained for a long time, and it can, under its own preconditions,

---

[3] cf. David Harvey, *Social Justice and the City*, London: Arnold, 1973.

call a new urban project into existence.[4] According to Roland Barthes, the 'myths of urbanism' draft a picture of the social whole. He writes that the myth does not deny things. On the contrary, its function is to talk about them. The myth makes things clean and innocent, it describes them as nature and eternity, and makes them transparent; which is not explanatory but ascertaining.[5]

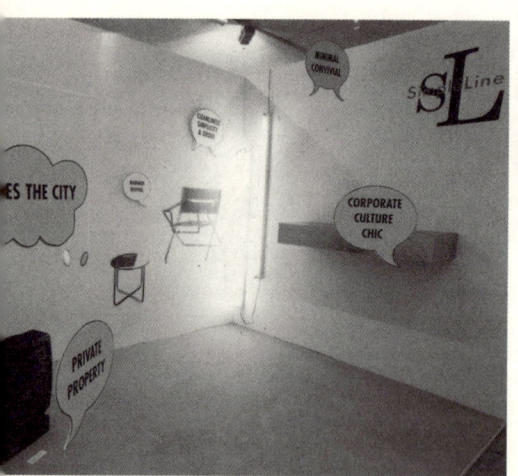

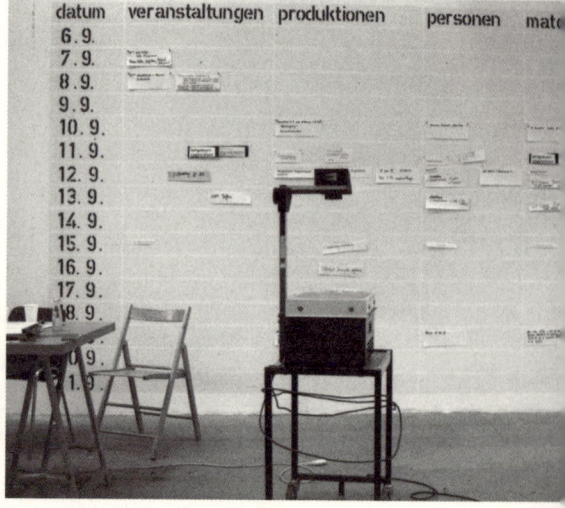

Sex & Space I, Shedhalle Zürich 1996
(Credit: Marion von Osten)

Marion von Osten and Pauline Boudry's installation, *Sex & Space II*, Forum Stadtpark, Graz 1997
(Credit: Johannes Gelliner)

It is against this background that the architect and designer Susanne Bachmann considered Godard's film *Deux ou trois choses que je sais d'elle* a valuable starting point for analysing what role images of women play in the portrayal of a city, what metaphors there are for representing femininity, and in what context they are used.[6] When describing or depicting a city,

---

4 See H. Hitz, R. Keil, U. Lehrer, K. Ronneberger, C. Schmid and R. Wolff (eds), *Capitales Fatales*, Zurich: Rotpunkt Verlag, 1995.

5 Roland Barthes, *Essais critiques*, Paris: Seuil, 1964.

6 This research by Susanne Bachmann resulted in a theoretical work as well as in a video production entitled *Cityquiz*. In this video the role of the emotionless, passive, female character is turned upside down: the artist uses it for the role of the quizmaster, who in fact introduced the gender stereotypes and metaphors.

one more often than not, it resorts to female images, ranging from the desirable woman offering herself to the explorer to the threatening, devouring and insatiable one.[7] In Godard's film, the traditional images of mother and whore are conflated. However, Godard places the emphasis of the film unmistakably on the role of the prostitute, using her as a metaphor that criticises the city. Although the whole of society is accused of being involved in prostitution, he demonstrates this only with woman. In particular, her commodity status is stressed: 'She appears to have no body, is fragmented, without any emotions or relationships.' It is traditional to use the 'female' metaphor of prostitution in order to demonstrate a negative process. The motif appears already in early representations of the city from the 1920s.

In the movie, the metropolis is represented as a place of conflict situated on the border between the bourgeois city centre and the proletarian suburbs. The new districts of entertainment are represented as dens of iniquity, and those spellbound by the large city and its temptations could not help ending up in a gutter as prostitutes, as 'Strassenmädchen'.[8]

The 'feminine' has often been allegorised as modernity through the figure of the prostitute, the commodity; as a motif it has been subjected to oblique and fragmented reference (or devaluation and disfigurement) that, nevertheless, embodies within the allowed interstices all the flux: uncertainty, exoticism, sensuousness, wealth of meaning, and 'otherness' that both the modern metropolis and the Orient in the nineteenth century unleashed. According to Christine Buci-Glucksman, 'the feminine constitutes one of the nineteenth century's "original historic forms"… The feminine becomes the inevitable sign of a new historic regime of seeing and "not-seeing", of representable and unrepresentable.'

---

[7] Pauline Boudry took up examples from the history of the genre of horror film and produced a video about such gender stereotypes. Horror films are quite instructive and useful when it comes to demonstrating gender difference as they depict fear in the face of the 'other' in a paradigmatic way. Quite often an affinity between the monster/alien and the woman is estab-lished: both are construed as biologically determined bodies that represent a threatening form of sexuality. On the other hand, 'woman' can be used to represent the unconscious as well as paranoid fear.

[8] See also Christine Boyer, 'Crimes in and of the City: The Femme Fatale as Urban Allegory' in *The 'Sex of Architecture*, eds. Diana Argest, Patricia Conway and Leslie Kanes Weisman. New York: Harry N. Abrams, 1996.

The red belt

Most of the city structures in which we live today emerged during the economic restructuring of the early nineteenth century. In the opinion of the French theorist Henri Lefebvre, the process of urbanisation is a central contradiction of capitalism. From the very beginning of industrialisation, the right to a dwelling place of the workers migrating from the countryside was linked to 'paid work' and 'rent'. To have a flat or, more precisely, a place to sleep was the main pre-condition to finding work. The living conditions of large groups of people were influenced by this interdependence of work, salary, and abode. This changed family life as well as class structure and therefore the gender situation and everyday life attitudes.

Nearly all contemporary observers described the process of 'urbanisation' as a shocking experience. The stage for these shocking experiences was the street, as can be gleaned from several narrations and descriptions of nineteenth-century city life. In 1844, Friedrich Engels, for example, wrote in *The Situation of the English Working Class* that the crowd in the street was already repulsive, and that people were appalled by it: 'Those hundreds of thousands of all classes pushing and running past each other… as if they had nothing to do with each other.' He went on to emphasise 'the individuals' brutal indifference and their emotionless isolation in the pursuit of their private interests', adding that 'this was more disgusting and painful the more crammed together people were in a small space'. For Engels, urban life was a paradigm for the 'alienation' of high capitalism, which he linked, among other things, to the social misery of the working-class districts. The new social order was mirrored in the segregation of urban settlements. On the city peripheries, settlements of migrant workers from the countryside became slums, due to low wages and high rent. This spatial organisation enabled the people in the bourgeois city centre to move about without encountering 'misery'.

It was this topographic 'misery' that made public health workers and reformers try to air, 'clean' and destroy 'these narrow alleyways and corners'. From 1848 until the Modern age, they wanted to do away with the 'prolific (red) belt'. The first large measures of city planning had socio-hygienic reasons. Besides gestures of charity, bourgeois life styles and attitudes were to be instilled into the working class by introducing new health rules, a sense of affluence, and moral codes.[9] Even Engel's descriptions of the metropolises of Paris and London contained conservative elements, which complained about the loss of agricultural identity in the greater family. 'Human nature' and its original system of relationships – the 'family' – were endangered by the city. Moreover, working-class districts were blamed for being too constricted and the cause of 'illnesses'.[10] City life itself was seen as 'subversive' to social relationships.[11]

In nineteenth-century literature encounters in the street were described without differentiation as encounters with the 'masses' or 'crowds'. The 'throng in the street' was perceived without difference of class, and public city life was compared

[9] At the beginning of the nineteenth century, women who worked in production often used to live on their own with their children, without men, in worker homes for women. In the patriarchal world-view this was equated with a coarsening of morals. Unquestioningly, women were appointed a place in a home, as for the bourgeois social politicians this was a way to keep the family under control. And male workers were afraid of a cut in wages (women's wages were half those of men), so for them this socio-political order was a way to push women out of production work, thus avoiding the loss of their own traditional privileges. This model of society had no place for female success, and introduced changes in dowry and inheritance law, making the man the provider for the woman, with the result that she was left without independent means. She was positioned inside the house, and as educator she mediated between state, school and institutions, and she guaranteed the preservation of the bio-political order. The man was positioned on the outside in 'public' life, and guaranteed steadily increasing prosperity and profit. These socio-political measures created a connection between public and private life as well as between city and family. It was during the new women's movement of the 1970s, when the private was required to be seen as political, that these problems were brought to the public conscience (see Susan Sontag, *Illness as Metaphor*, New York: Farrar, Straus & Giroux, 1978).

[10] 'The 'cancer' metaphor extends the theme of the rejection of the city. Before being construed as a carcinogenic environment in the literal sense of the word, the city was described as a cancer – a place of abnormal growth. Tuberculosis and the real or alleged threat of this disease perpetuated among slum improvement and living space movements of the late nineteenth and early twentieth century generated the assumption that slum dwellings were breeding grounds for tuberculosis. The 1950s saw the transition from tuberculosis to cancer in the rhetoric of planning and house building organisations. The use of the term 'invasion' to describe the move of non-white and destitute citizens to middle-class neighbourhoods constitutes another metaphor, borrowed from both cancer and war; indeed, in this instance the discourses cross. In *The Living City* (1958) Frank Lloyd Wright compared the city of former times, the healthy organism (the city was not yet seen as malignant then), with the modern city. For him, to look at the city on the map, divided into grid squares, is like looking at 'the cross-section of a tumour' (cf. Susan Sontag, *Illness as Metaphor*).

[11] The French statesman Haussmann, responsible, among other things, for urban planning in Paris during the Second Republic, saw in the urban geography a subversive potential for revolt. He planned a considerable number of boulevards, which were also built (e.g. the Champs-Elysées), but he first and foremost intended them for military purposes. Quite soon, however, he noticed that when they were used for revolts the broad avenues could not be controlled very efficiently.

to the production process of 'a big machine'. In his *Zentralpark* (Central Park), Walter Benjamin demonstrated this change of perception, mainly by using the work of Baudelaire. The large city turns into an amalgam of kaleidoscopic, simultaneous experiences for the stroller. He surrenders to the street, 'as to a woman'. This male voyeur, the *flâneur*, perceives difference only by sexualising the street, by perceiving the passing woman or prostitute as 'other'.[12]

Although there were wide-reaching socio-political measures to reduce women to their reproductive role, it was the capitalist city that pushed women more and more into public space, acting as workers and employees, as dancers, *femmes nouvelles*, prostitutes and consumers. The bourgeois male observer or stroller, according to his sexist logic, saw this as a temptation. However, it also meant danger and the dissolution of existing social relationships. The transformation of social relations and identities led to a description of the city as 'dangerous', and a cause of moral decline. In the centre of this transformation were socio-political and reactionary discussions of the 'other' in comparison with the so-called 'normal'. The focus of these sexist and racist attributions of the nineteenth century was prostitutes, lesbian women and gay men, Jews and blacks. Female bodies were equated with the sexual woman who sells her body, as syphilitic and 'unclean'. In his book, *Difference and Pathology: Stereotypes of Sexuality, Race and Madness*, Sander Gilman described how these constructions of the 'other' in the nineteenth-century popular imagination have a common ground with the racist attributions ascribed to Jews in fascist Germany. In the bourgeois novel of the nineteenth century, women – above all, proletarian women – were associated with the dangers and 'diseases' of city life.[13]

---

[12] Walter Benjamin, *Selected Writings: Vol.4: 1938-40*, Cambridge, MA: Harvard University Press, 2003.
[13] According to Gilman: 'In both *Josefine Mutzenbacher* and the *Victorian child pornography*, the young girl is depicted as an active participant, if not indeed as the originator of the sexual act. The availability of the young girl as sexual object is highlighted. It is in fact not just any girl, but, at least in Vienna, the proletarian one especially. Her sexuality is the sexuality of the 'other'. The distance between the self and the other is the same as the distance between the races. For Jeffer, it was the black man's sexuality as other that turned woman into the sexual object par excellence; for Freud it is the lower class, proletarian woman who constitutes the main representative of corrupting sensuality'. (Sander Gilman, *Difference and Pathology: Stereotypes of Sexuality, Race and Madness*, Ithaca, NY: Cornell University Press, 1985)

Yvonne Doderer and Cora Schafer's installation,
Sex & Space II, Forum Stadtpark, Graz 1997

Bourgeois theorists saw the city as an 'artificial' order, compared to a 'natural' one and the 'city planner' as disciplinary authority was invented. This new authority was installed in order to control the 'chaos', separating public from private space, and the centre from the periphery by architectural means. This 'anti'-city movement believed that it was mainly the dissolution of family structures, and the change in national identity and traditional gender roles that threatened society, and not the growing city structures or the increasingly delicate hierarchy of exploitation. According to Jane Jacobs, in *The Life and Death of the Large American Cities*, this anti-urban, paranoid view of the city, the fear of the street and the 'lurking danger of the large city', was the cause of modern planning mythologies.[14] Architectural reform, Garden City movements and current suburban settlements were brought about by the fears of a 'decomposing' city life. This caused Le Corbusier's Cité Radieuse, through the Charter of Athens, to reverberate in the popular imaginary, as it still does today, and resulted in the concept of the 'Neue Einfachheit' (new simplicity) for Berlin, the newly declared capital of Germany, as well as for other European cities. These concepts equated city life with the mere act of consumption and security.

[14] Jane Jacobs, *The Death and Life of Great American Cities*, New York: Random House and Vintage Books, 1961

The construction of the 'other' is tightly linked to these projections onto public space. The classification of gender and race determines the perception and order of public space in architecture and city discourse, as well as in texts and other representations of urban space. On the occasion of the project *Sex & Space,* the artists and theorists intended to look at the systems of representation of 'genderised' space and the related policy of segregation. When analysing the systems of representation of urban space, we wanted to study, on the one hand, the appearance of an image, or production of a myth, and, on the other, to read these narrations of spatial concepts as social power structures and the systems of spatial representation as the representation of social relationships.

His story?

In the common discourse on architecture and planning, analyses of power relations and systems of representation are missing. This is the reason why it is still commonplace for the creation of a building and an aesthetically formal paradigm to be given a mandate for social moulding. This 'agreed system', stemming from Modernism, negotiates the functions and areas of responsibility for architects, planners, politicians and economists. City planning authorities and building commissions use this common practice in order to decide on either access or exclusion, hiding behind the apparent neutrality of aesthetic paradigms.

In the texts of Modernist architects (Gropius, Meyer, Le Corbusier, etc.), compiled in Konrad Wünsche's book *The Bauhaus or the Wish* (1992)[15] in order to organise the World, we come across the indomitable belief that social conditions can be changed when designing the appropriate physical environment (through interior design, architecture and urban planning). Society was described therein as a homogeneous unity and this formed the basis for the democratic and social tasks assigned to architects and designers after the First World War. This task was seen as pedagogical, to improve social and

---

15 Konrad Wünsche, *Bauhaus: Versuche, das Leben zu ordnen (The Bauhaus or the Wish)*, Berlin: Wagenbach Verlag, 1992.

moral conditions via the construction of a 'new man' and to improve spatial living conditions by allowing more light and air. The positive effects of this movement will not be denied, but it is also important for the understanding of architectonic space, which is still interpreted as universal and democratic, to see that a 'socialist' design for the people – a functional, pure vocabulary of forms for 'all' – was being developed. This was meant, on the one hand, to liberate society from bourgeois representational pomp and overly confined worker districts, but on the other, it served the architects and designers as a means of stylising themselves as all-embracing artistic geniuses and the new intellectual elite. His 'social mission' allowed this new type of designer to distance himself from engineers and masons and to declare 're-designing' as the conditional basis for a 'new' social order. The Modernist architect was thus born; his gender, male.

'Architecture' and 'masculinity', two apparently unrelated discursive practices, are seen to operate reciprocally in the remarkable opening scene of the movie *The Fountainhead* by King Vidor, as Joel Sanders points out in the exhibition catalogue 'Stud, Architecture and Masculinity'. The movie exploits building metaphors to articulate the theme of 'man-worship', while the portrait of Howard Roark, the architect as 'creator', sanctifies architectural doctrine. In the novel's central dramatic scene, the courtroom scene in which Roark is being tried for blowing up one of his own buildings during construction, Rand's uncompromising male idealist defends the principles of modern architecture with arguments comparing built structures to masculine virtue, claiming that buildings have integrity, just like men. Roark's narcissistic proclamation echoes the words of Western architects and theorists from Vitruvius to Le Corbusier who, in their attempt to locate and fix architecture's underlying principles in a vision of trans-historical nature, recruit masculinity to justify practice. Rand's architecture of masculinity offers one of the most dramatic, although certainly not the earliest, renditions of the notion that buildings derive from the human form itself, specifically from the unity, scale, and proportions of the male body.

One aspect of this modern construction is the interior white wall and the status of this supposedly pure non-colour. It combines purity and at the same time an ability to define what is not pure, what is 'other'. The lecture performance by Lukas Duwenhägger and Dominic Eichler on the occasion of the project *Sex & Space* interpreted the white wall as a result of purification, control and discipline fantasies of modernist architects, which represent a latent and homophobic panic of a 'feminisation' of culture. The project of the artists Josef Strau and Michael Zinganel aimed to put the utopia of Modernist architects and city planners in a historical context. Their lectures shed light on the question of which spatial and ideological conditions reinforce gender, class and race relations in Modernist building concepts, even though these concepts claimed to be free of any such ideologies. Garden Cities, co-operative settlements and one-kitchen houses are examples of building projects with a progressive land-price policy. They are, nevertheless, simulated communities in which racist exclusions and entry regulations of nationality or classical (heterosexual) family structures persist.[16]

The model of the Frankfurt Kitchen by architect Margarete Schütte-Lihotzky, an architectural reform project for social apartment buildings in the 1920s, was the starting point for the work of Berlin artist Cornelia Schmidt-Bleek. Schütte-Lihotzky developed the first feminist design project that aimed at facilitating domestic chores by reducing distances and introducing functional design. Realising this project, however, did not mean that women were able to rid themselves of the traditional duty of having to do the housework. The kitchen, rationalised by methods of business management, became a focus for public discussions, resulting in a wave of technological fantasies. Technologies are, to this day, produced by women at a low price and consumed by the same

---

16 In the same way, Yvonne Doderer worked on the relationship between lesbian autonomous spaces, their contribution to the constitution of lesbian identity and culture, and their possible representation. Large photographs depicting lesbian women of her community had been hung on the walls of the Forum Stadtpark, a construction built in the purest Modernist style. These very striking representations of lesbians, which were related to two documentaries and a critical text about lesbian women in Germany, raised the question of how radical feminist and lesbian positions could be revitalised, that is outside the lesbian chic phenomenon, which is of no social or political importance to lesbian women whatsoever.

Sex & Space: space / gender / economy 229

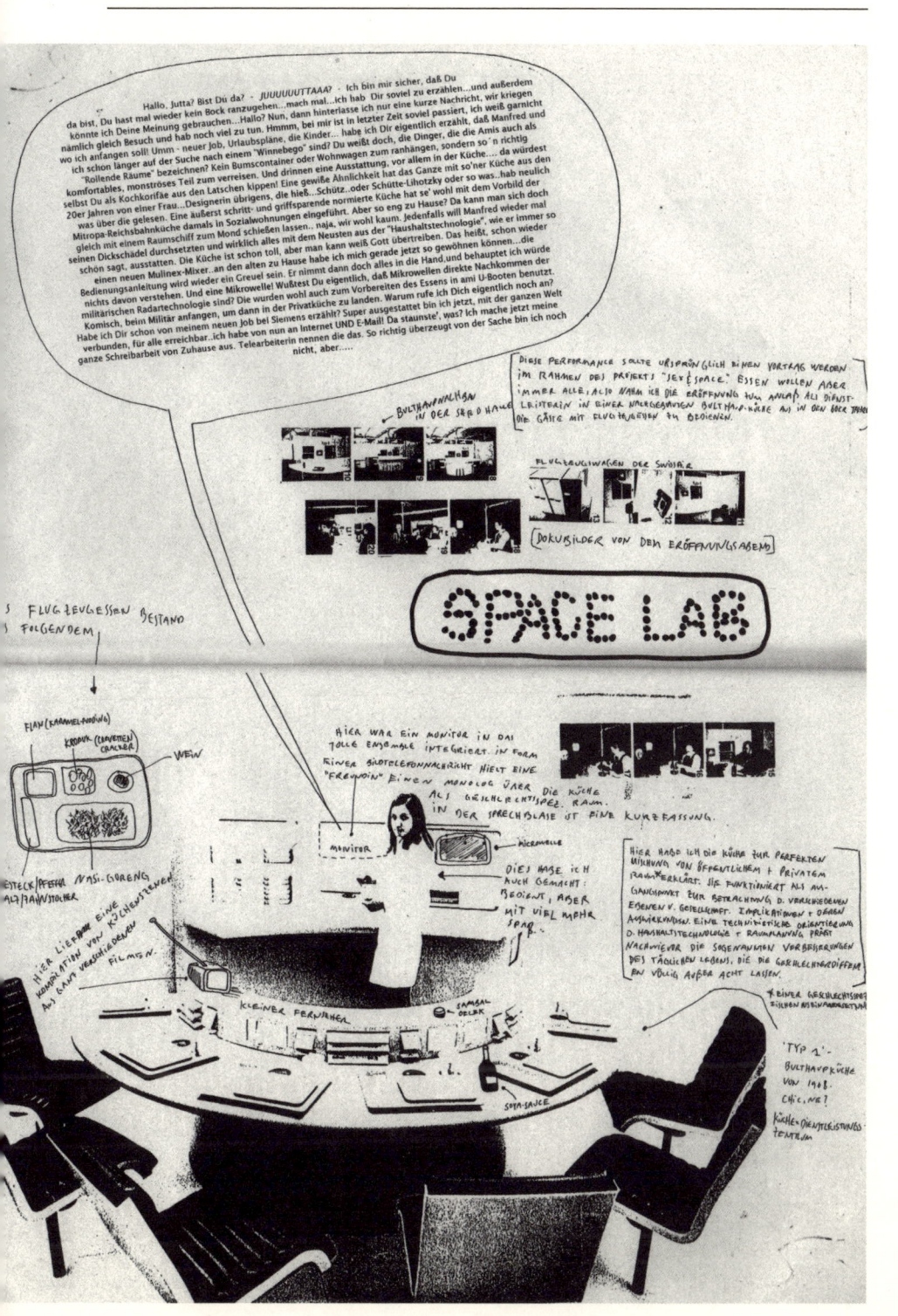

Space Lab in *Sex & Space I* fanzine

at a high one.[17] Today the hype surrounding new communication technologies such as 'Home-Offices' and 'Smart Houses' results in a renewed promotion of traditional role models, and 'woman' as moderniser is declared an economical factor not only in advertising.[18]

Her job?

During our analyses, we found that even left-wing city sociologists had hardly ever looked at the question of traditional gender-specific division of labour and the way it determines spatial organisation, let alone the question of how people are affected by the latter. Therefore, we focused on the history of female socialisation, and how this has been influenced by the traditional sphere of male salary biographies, i.e. their work, which is socially acceptable and has its designated place in public life, while reproduction work has neither a financial nor a symbolic place in public space. Modern cities are traditionally divided into 'masculine' city centres of production, based on male socially valued salary biographies, and 'feminine' suburban zones, associated with women's unpaid work, unrecognised as a part of the social sphere. This spatial division is caused by the law of labour division in our society, where one half of the work necessary for the functioning of society is valued and remunerated, and the second half is unpaid and not really considered work at all. The construction of such separate zones conditions the organisation of work, the norms of building and living, social roles, and the possibilities of moving in public space. In post-Fordist city structures, the division into leisure, workplace and housework is, at least partly, not as clear and women see themselves associated with new attributions.

---

[17] See Cynthia Cockburn and Ruza Fürst-Dilic (eds), *Bringing Technology Home: Gender and Technology in a Changing Europe*, Buckingham: Open University Press, 1994.

[18] A commercial run by German Telecom at the time of the exhibition illustrated this interrelation. In this clip a young woman, dynamic journalist and mother, is depicted using the internet to work for her newspaper from the comfort of her home, tending her baby at the same time. How is she able to manage the two things simultaneously, family and job? The answer, so the message of the commercial runs, lies in the use of new technologies, the transition to flexible working hours and the conflation of private space and workplace. Only these innovations, the commercial implies, enable her and her company to solve the problem of the extra workload smoothly and efficiently. But in fact, she has returned to the 'workplace' traditionally assigned to her.

The decision to deal with 'work' in connection with the project *Sex & Space* was amply justified by the fact that work still figures quite prominently, if not increasingly so, in an individual's life, and moreover, lends itself to reflecting and analysing the question of gender in all its various aspects, including the possible chances and perils of change.[19] The complex starting point of *Sex & Space* meant that the analysis of 'work' was not to be approached one-dimensionally or in a way that would rely too heavily on a central emphasis;

Sex & Space I, Shedhalle Zürich 1996
(Credit: Rote Fabrik)

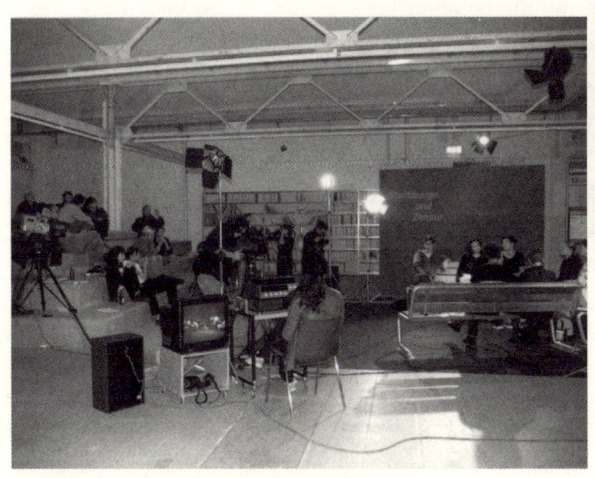

'City Design and Censorship', panel discussion at *Sex & Space I*, Shedhalle Zürich, 1996 (Credit: Rote Fabrik)

instead it seemed more rewarding to follow up the interrelation and interdependence of the various aspects. Thus, it was not only from a sociological viewpoint that 'work' was seen as fundamental for the constitution of the human self. Questions such as how geographies manifest themselves in conditions of work, where and when gender-specific

---

19 'If we accept... the sociological premise that in our society life options are first and foremost determined by the kind of work a person performs, the general assumption of a continual improvement of the social standing of woman has to be qualified.' See Ursula Rabe-Kleberg, *Careers for Women: On the Segmentation of Professions*, Bielefeld: Kleine Verlag, 1987.

changes occur in the workplace and whether and when these trends are counteracted by, although traditional, more differentiated stereotypes appear.[20]

After dealing with this complex theme in a rather theoretical way, Brigitta Kuster and Rachel Mader decided to tackle questions concerning work and the specific situation of women involved in it by means of interviews with 'atypical' employed women. 'Atypical' insofar as these women did not show the criteria of a classic male salary biography[21] and their lives were characterised by a plethora of breaks and inconsistencies. With hindsight, this choice, seemingly little reflected, proved quite illuminating as it inevitably involved questions of terminology and evaluation. The six women interviewed were all in a situation that required them to work. Although they were hardly identifying with their gainful employment, their 'other' lives proved all the more varied and interesting. They were engaged in various activities of a cultural, social or athletic nature, which, however, were badly paid or not paid at all. Nevertheless, all the women described this part of their lives as both more important and more satisfying.[22]

Finally, the issues dealt with in the project showed up in our own microstructure and we came face to face with the complexities as well as the possible interventions of

---

20 For her initial investigation the artist and theorist Brigitta Kuster decided on the company 'Sol'. 'Sol' stands for a typical example of new spatial and organisational working conditions. The building has a non-hierarchical open plan office structure, desk sharing and on-line home working are part of the company's work policy, and furthermore, the office has been given an exaggerated Laura Ashley look. If the so-called female biographies are anything to go by, this is in fact the perfect place for women to work in. But when, on the one hand, it appears that women prefer to work under flexible working conditions, this also means that educational work and housework traditionally associated with women are not called into question but taken for granted. The question whether the feminisation of work should be regarded as progress or rather regression is still much debated by feminists. Part-time work is a further example of the ambivalence of 'feminised' working conditions. Ulla Kilchenmann, in her extensive analysis of the issue, comes to the conclusion that employers soon learned how to use part-time work in their favour. See Ulla Kilchenmann, *Flexibel oder flexibilisiert? Chancen und Fallen der Teilzeitarbeit von Frauen*, Zürich: eFeF-Verlag. 1992.

21 Frigga Haugg is one of the many authors who, since the early 1980s, have been analysing the construction of a traditional male salary biography in connection with criteria such as Neo-liberalism, political facts, etc. She stresses not only the fact that a 'male salary biography' was – and indeed still is – dependent on the separation of public and private as well as the wife and mother doing the housework for free, but also that such a term omits all 'caring, reproductive and such activities that show responsibility for the self, one's fellow human beings and the rest of nature that, with respect to one's own life, ought to be treated in a caring way'. See Frigga Haug, 'The Neoliberal Project, the Male Definition of Work and the Long Overdue Revision of the Contract between the Genders', *The Argument*, 217 (1996).

'alternative models of employment'. During a collective preliminary work session with the artists, we discussed the form and content of their contribution, which had to fit a given framework, as well as the design of the exhibition display and the scheduled events and workshops. This not only implied a potential increase in the people participating, but also meant more work, thus generating tensions especially when decisions were asked for, payments had to be justified or responsibilities needed to be taken.[23]

In the second part of the project, this time in Graz, in conjunction with an extended team, we laid the emphasis of the previous weekend's events on the question of whether the current discussion about new working conditions confirms Western hegemonial values and, if so, how they manifest themselves. The discussions by feminist theoreticians who felt, on the one hand, ambivalent about the transition to

22 One of the interviewees – Lilo – aged about 40, spent 40 per cent of her time working as a secretary, 10 per cent as a night porter and the rest she dedicated to the cinema at an alternative cultural centre. This 'leisure activity' was the one she could most identify with emotionally and intellectually. Although Lilo was, to a great extent, the only one responsible for her then work situation, she was experiencing her condition as ambivalent and partly unsatisfactory. Due to the social hegemonial division of gainful employment and leisure time, people were not able to judge her salary biography adequately: no one considered her to be primarily a film expert, and she hardly ever introduced herself to other people as such. The fact that we stage-managed our performance in an office-like setting within the exhibition hall should be understood as a critique of the selfsame structural logic of the term 'work': badly paid or unpaid work hardly enjoys any social prestige at all and is more often than not received as less 'serious'. In a discussion taking the form of a workshop, we introduced texts and surveys that dealt with the gendering of new employment models. We also used the workshop form in order to deal with the fact that gender-specific attributions are subject to constant change and their deconstruction has to be based on the careful observation and description of shifts and displacements. So in the interview with Lisa, a single mother, we found that although she is a trained bookseller, she works in the accountancy department of a library. Deeming her job unsatisfactory, she intended to give it up as soon as possible but refrained from doing so on account of the social security and agreeable social environment the job offered. This decision might be rashly dismissed as a typically female attitude, since women are traditionally seen as the ones responsible for all inter-human relationships. It is important to reject this interpretation, however, because Lisa's situation, more than anything, shows what professional opportunities are closed to women due to their specific situations: a single mother cannot afford to be employed under insecure conditions of work, as she cannot rely on someone to cover her basic needs and to take care of her reproductive tasks (e.g. child care).

23 A third interviewee – Martine Anderfuhren – took part in *Sex & Space* in her function as artist. Working as a consultant in a video library, she was at the same time doing a postgraduate course at an art school. For her, participating in *Sex & Space* meant, on the one hand, artistic stimulation and the satisfaction of her interest in the creation of an art-specific network that would secure her future, but, on the other, it was also very time-consuming and meant that she had to manage her time carefully so as to be able to fit in her two additional jobs. She too, therefore, was doing a work she in fact appreciated more, but was only rated as badly paid leisure time activity – conditions that are, in the final analysis, caused by the classic division of paid and unpaid work and the gender-specific spheres connected therewith.

flexible working hours, but on the other, criticised long-time biographies of middle-class European men, were taken as a basis for gaining new insights into international forms of division of labour and its spatial and gender-specific effects. The situation of women migrants in Austria, who are not allowed to work due to the principle of allowing families to remain united, and are thus forced into traditional dependence, served as a touchstone for our own political concepts. The weekend, therefore, was planned as a forum, where various feminist and anti-racial approaches were not only to be introduced but also analysed with a view to their strategic and performative potentials.

Cartographies of fear

Compared to the industrial age, cities today increasingly serve as centres of knowledge, places where information is produced and disseminated. Under the pressure of the international free trade agreements, the cities of the 1990s have been treated like capitalist enterprises. A modern city management handles the location advantage of their city, its qualities and infrastructure, as saleable goods, and puts them on the market accordingly. The enterprise city no longer orientates itself to the needs of its inhabitants but to those of the tourist trade and investors. Nowadays, city marketing increasingly includes open-air festivals and large, spectacular exhibitions. Within this context, art and architecture have become international economic and prestige factors of the 1990s. In the process of a post-Fordist restructuring of the city (flexibilisation, deregulation and lean management), new museums and theatre buildings have emerged along with the transformation of society, making culture an important commodity which has helped to give cities a more attractive locational factor.

City managements compete internationally, and resort to lavish interventions in the city structure, in order to secure locational advantages within the global market. Cultural models are needed to justify these interventions. New museum buildings, and the new trendy bars and consumer areas that come with them, bring a better infrastructure for traffic and communication, along with the 'culturalisation' and 'increase in value' of existing districts. The American sociologist, Saskia Sassen, describes the importance of *Global Cities* with regard to their role as important switchboards for deregulation and restructuring processes. The economic concentration of power in a few large cities in the world increases not only the well-paid service sector, of mostly white Western men, but also a large informal sector (cleaning, supply service) of female workers and migrants. Nowadays, the urban centres are seeing a trend towards the development of hegemonial social claims by a white elite working in the services sector. There is an attempt at regulating their gaining a foothold not only via state boundaries and other controls, but also by means of a massive discourse on security and an ongoing privatising and commercialising process of urban space.

Since the beginning of the 1990s, the city policy of 'cleaning the city centres', with an increased police presence, has led to a repressive drawing-up of borders. Those people who cannot be integrated into the normative logic of a bourgeois life have been represented by the media as 'pathological' or 'criminal', and executive measures have dismissed them from the public 'eye' of bourgeois life. Fears multiplied by the media, of the supposed creation of new slums, and of foreign infiltration in various districts of the city, have resulted in an increasing everyday level of racism as well as in a paranoid boom in private security services. The construction of the 'other' is tightly linked to projections onto public space. Under political and popular pressure, and supported by a large section of the media, the city governments adopted a repressive drugs policy at the beginning of the 1990s.

In 1992, for example, this discourse led to the eviction of junkies and drug dealers from Platzspitz, a park in Zurich, just behind the main railway station, which had become Europe's largest open drug market, known worldwide as 'needle park'. As a result, junkies and dealers spilled onto the adjacent residential neighbourhood of the district. The implementation of the above-mentioned policy had not only a negative effect on this neighbourhood, but also worsened the social and political climate of Zurich. The government, and even large parts of the urban-cultural population of the city, adopted a racist discourse accusing 'illegal immigrants' and 'criminal refugees' of being responsible for the drug problems in Zurich. Under the pretext of fighting the dealers, a whole system of repression was created, ranging from new national laws directed against refugees, to the construction of a whole series of new prisons and camps to house the dealers, illegal immigrants and junkies. The banishment of drug users and dealers from public space was merely the beginning of a policy that tried to keep all inner-city public space exclusively for the 'integrated' part of the city population. Subsequently, quite a few attempts were made to exclude more segments of the population from occupying public space. This time it involved certain categories of immigrants, women working in the sex business (a large part of them are illegal immigrants), tramps, etc., in fact, everyone considered to be a non-consumer. These drastic measures were accompanied by a boom in the privatisation of zones restricted to consumers, and by an increase in private security services. A further remarkable phenomenon was the emergence of expressions with racist overtones to describe certain parts of Zurich, a city considered to be one of the cleanest and safest in the world. Slogans such as 'Zurich is turning into a slum' were used.

In connection with such polarising tendencies, the discussion about the 'right' social mixture for a district is fast gaining importance in Germany as well as Switzerland. A policy that favours a certain percentage of foreigners spread evenly throughout the entire city is said to generate a 'healthy' ethnic make-up of the population and ensure the integration of foreigners; a spatial concentration of immigrants, however, will lead to parallel societies, which may give rise to fear and xenophobia. Due to this kind of reasoning, for some years now so-called immigrant ghettoes have come to be the focus of the media, politicians and experts. The images reinforced by such attributions, however, do not reproduce the object they refer to; instead they mirror the specific interests and conditions of those responsible for them. By means of collective symbols, characterised by the tropes of catastrophe, violence and social pathology, the ghetto as 'the other' is made into the imagined normality of the middle class. As specific constructions of reality, the discourses on ghettoisation constitute procedures of either inclusion or exclusion and therefore must be understood as social practice: The allegedly legal drawing up of a border between social inclusion or exclusion, between the 'normal' population and the inhabitants of the ghetto, is facilitated by attributions of blame for deepening social and cultural rifts and for violence directed at civil urbanity. For the parts of the population already discriminated against, this stigmatising definition of one's own district by the social centre is equally relevant with respect to the process of marginalisation as their actual social situation.[24]

Innerstadt Aktion, newspaper, June 1997

From a feminist point of view, public space has been treated as a space of female fear for years. Although women mainly suffer violence in private spaces, i.e. their own home (over 80 per cent), mostly in marriage, in feminist publications fear has conventionally been defined as women's fear of male violence

[24] Apart from readings and films, the activist group 'Innerstadt Action Zurich' also organised various actions and demonstrations against this policy. During the second part of *Sex & Space* in Graz, some members of this group tried to analyse the role the media had played in making the policy more attractive. One problem they had to deal with here lay in the question of how articles and illustrations could be deconstructed, without actually showing the images that support the city's restrictive policy. For the exhibition part, they produced a radio programme criticising the main-stream articles. A rather ridiculous picture of two city workers scrubbing off graffiti was hung on a wall and served as a setting for the programme.

occurring in public spaces, thus reiterating, on the one hand, the assumption of men being the culprits and women their victims, and, on the other, the idea that public space is associated with danger, to be avoided (by women), or to be controlled (by socially competent women). This notion of public space therefore reinforces traditional gender roles and the concept of 'difference', and is confirmed by the way public spaces are created and by their spatial consequences. Examples of so-called good 'solutions' to try and make public places safe by using 'order', 'light' and 'cleanliness', have claimed to be able to 'prevent violence', but they in fact have the opposite effect. It is the subtle colonisation of public space by the respectability of a dominant white bourgeoisie that produces, not without good reason, aggression and vandalism. It is not my wish to dismiss the approach of feminist planners, but this is a paradigm for the anti-urban, disciplinary models, which infiltrate even emancipatory city discourses. In the 1990s, the safety discussions concerning public space resulted, even in feminist approaches, in the production of 'fear' maps where city quarters, in which 'unpleasant subjects' were to be expected, were marked with a star. The Swiss geographer, Nicole Stolz, confirmed the assumption that most of the fears expressed by women she had asked during her survey were produced by the media.[25] She found that most places that were designated as 'dangerous' by the women who were interviewed had previously been described by the press as places connected with drug addicts, youth, and the racist stereotype of 'foreign' criminals.[26]

Mary McLeod, an American architectural historian, points out that although since the Second World War the concept of 'other' has had a powerful influence on Third World political and cultural theorists (from Frantz Fanon to Edward Said),

[25] This aspect was taken up and dealt with in a video by Katja Reichhard made in connection with a subsequent project on urban polarising tendencies, entitled *Baustop Randstadt*, staged in Berlin, in 1998. The video tries to demonstrate how, due to a discursive correlation of neglect, drug addiction, criminality, high percentage of immigrants and destitution, certain districts are subject to being labelled ghettoes by the mass media.

[26] The discussion of public space as a space of female fear, as conducted in publications by Swiss and Austrian planners, led to the idea for a public performance: in Geneva, eight women spent the night of 13 July in beds they had moved outdoors, into the public realm. The artist Martine Anderfuhren, herself from Geneva, has created a screen saver that takes up the subject of this performance.

most contemporary theorists have never attempted to look at the 'other' from the perspective of the 'other'. In poststructuralist philosophy and literary criticism, a major claim for political validity is the notion of dismantling European logocentrism. Yet, despite this embrace of the 'other' in some of its theoretical sources, contemporary theory in architecture would appear to posit a notion of the 'other' of a white Western male cultural elite. Instead of investigating the question of what the avant-garde's desire for 'other' is, architects and theorists would be better asking what the desires of those multiple 'others' – real flesh-and-blood 'others' – are.

The process of post-Fordist reorganisation, currently taking place, is generating new lines of conflict along the boundaries of gender and race. Female migrants and refugees working in Western cities for very little money and having no legal status are just as much part of these developments as are the new employment models that allow 'gender-specific qualities' or new forms of exclusion and their racial consensus. On the occasion of public talks with planners, activists and architects, we discussed and criticised Western-dominated feminist perspectives, which have become the ruling logic for popular ideas of expulsion in West European cities.

The bipolar opposites of inclusion and exclusion, public and private, respectively, are expressions of a power structure that has always favoured very specific sections of the population with regard to possible forms of behaviour and representation. Contemporary feminist practice and theory should be capable of realizing that power is not just patriarchal power against women, but that power is subject to constant production, reproduction and attribution processes. As Mary McLeod emphasises in 'Everyday and Other Spaces':

Difference is experienced differently, at different times, in different cultures, by different people. The point is not to just recognize difference, but *all kinds of difference*.[27]

---

[27] See Mary McLeod, 'Everyday and Other Spaces', in D. Coleman, E. Danze and C. Henderson (eds), *Feminism and Architecture*, Princeton, NJ: Princeton Architectural Press, 1996.

Chapter 15

Arthur Kroker and Marilouise Kroker

Niran Abbas

Claudia Springer

REFIGURING
DIS/EMBODIME

Dawn Dietrich

Laura Gurak; Katherine N. Hayles

Sherry Turkle

Ernest Larsen

Marie O'Mahony

Gilles Deleuze and Félix Guattari

Sherry Turkle and Seymour Papert

Bernadette Wegenstein

Donna Haraway

Chapter 15 Niran Abbas

Elizabeth Grosz

Patricia Clough
Mark Hansen

ENTS

Marcos Novak

Andrew Murphie and John Potts

Francisco J. Varela, Evan Thompson and Eleanor Rosch; Andreas Huyssen
Marie-Luise Angerer
Allucquère Roseanne Stone
Margaret Morse

In *Discipline and Punishment*, Foucault (1975) shows us how architecture can embody relations of power and how relations of power themselves constitute architecture. Gender relations, like all power relations, are organised in time and space. The architectural arrangements of homes, workplaces, schools and other institutional settings, as well as geographical ordering in the broader sense, are simultaneously gendered spaces. This is an old story. Practices of gender, exclusion (confinement, segregation) by physical means have a long history. Already in the fifteenth century, Léon Battista Alberti was describing the home as masculine territory, a space for enhancing masculine status and power. This pattern is reproduced in countless ways in building design industry and technology, both public and private.

The industrial-age tendency to apply gendered metaphors continues in the electronic age but is complicated by the computer's ambiguous relationship to gender. The design of a computer does not immediately evoke either male or female attributes; if anything, it presents a bland and asexual surface. But the urge to assign a gender and therefore *reinstate stereotypes* to technological and other cultural artefacts persists with controversial effects.[1] Consequently, in an attempt to masculinise products, manufacturers refer to their computers' power and strength; femininity equals small size, fluid, internal workings and quiet functioning, and the ability to absorb the user's ego in an emphatic bond. Still others celebrate the computer's supposed gender-bending capabilities that allow users to celebrate their on-line personae: men can become women and vice versa. What the various contradictory gendered metaphors make apparent is that computer discourses incorporate current cultural debates over gender roles.

This chapter provides a glimpse of some of the theoretical aspects of the 'hyper-texted' body (to borrow Anne Balsamo's term) on the electronic regulated frontier. In one sense, this frontier is an imaginary construction that identifies a horizon

---

[1] Children's toys are filled with gender bias. A case in point is Mattel's computer's designed for boys and girls: the pink and flower-power 'Barbie Computer designed for girls' and the blue box painted with a bright yellow flame called the 'Hot Wheels Computer designed for boys'. The advertisement ran in many national newspapers and on the Internet during late fall and winter 1999. Citation in Laura Gurak, *Cyberliteracy*, New Haven, CT: Yale University Press, 2001, p.76.

of contemporary cultural thought. But in another sense it is a real space on the fringe of mainstream culture: the 'electronic frontier' names the space of information exchange that already exists in the flow of databases, telephone and fibre-optic networks, computer memory, and other parts of electronic networking services from which to construct a totalising vision of space. While print and radio tell, and stage and film show, cyberspace embodies and creates an immediacy or transparency of the medium – a dream that André Bazin had of total cinema. The coupling of the body and space at this particular moment in our coevolution of technology has 'suspended the framing function performed by the (preconstituted) technical image'[2] and has empowered the body into a realm that Marcos Novak imagined as the domain of *liquid architecture*.

Just as we can use an array of pixels to create any image we please within the confines of a screen, or a three-dimensional array of voxels to create any form within the confines of an overall volume, so we can create sense-shape with an array or volume of appropriate sensels. Such a shape would be exact, but invisible, a region of activated, hypersensitive space.[3]

Allucquère Roseanne (Sandy) Stone argues that cyberspace and computer-mediated communication involve the decoupling of the subject from the body. Stone complicates the artificial intelligence (AI) narrative of disembodiment by claiming that this very de-coupling makes cyberspace 'a locus of intense desire for refigured embodiment'. At the same time, she suggests that this refiguration is often purchased at the price of 'freedom *from* the body', and that cyberspace therefore often appears only as a concretisation of the 'psychoanalytically framed desire of the male'.[4]

The emergence of cyberspace or mind–space technology reframes the question of how experiences are encoded as

---

[2] Mark Hansen, 'Wearable Space', *Configurations*, 10 (2) (Spring 2002): 322.
[3] Marcos Novak, 'Eversion: Brushing against Avatars, Aliens and Angels', AD: *Hypersurface II* (1997): 72–6.
[4] Allucquère Roseanne Stone, 'Split Subjects', *Configurations*, 2 (Winter 1994): 109, 107. In science fiction, Candas Jane Dorsey's story '(Learning About) Machine Sex', in *Machine Sex and Other Stories*, Victoria, BC: Porcepic Books, 1988, offers a similar feminist critique of men's attitudes toward interactive computer technologies and indirectly critiques the genre of cyberpunk fiction.

'interior' and 'exterior'. The collapse between the user and the screen questions the experience of embodiment (an exteriorisation of mind).[5] AI researcher, Hans Moravec, describes how one day 'uploading' of human mental functions will be transferred onto computer software in a process called 'transmigration'. In this narrative, the mind not only takes on qualities associated with the body, but the mind takes over those qualities supposedly making bodies obsolete. For some, the concept of obsolescence for humans becomes in practice a form of social Darwinism: the survival of those who have the economic means to finance their continued existence.[6] However, there is another story about the collapse of the mind-body dualism and its outcomes; what disappears is not material bodies but an abstract notion of the body as the naturalising ground of unitary and universalising notion of the self. The disappearance of the body is then followed by a reconstruction of embodiment. A reconstruction known as the posthuman.

We are, then, never disembodied. Simulated worlds can exist for us only because we can perceive them through the techno-apparatus of our body spliced into the cybernetic circuit.[7] We feel we are in a virtual space only because our senses tell us we are – through our eyes, our hands, and our ears. Even our haptic senses pick up the internal arrangement of our bodies, subject to things like movement or gravity, to tell us where we are in a virtual space. However, it is also true to say that the body will never be *the body* again. Since the 1990s, the inner body has been recorded and 'externalised' through new technologies – such as the *Visible Human Project* juxtaposing the Visible man and Woman as 'digital Adam and Eve' and the *Human Genome Project*. Katherine Hayles points to a possible shift from a culture based upon the presence or absence of bodies to society based upon patterns or randomness in information – a 'dynamic flux from which both the body and embodiment emerge'.[8]

5 Arthur Korker and Marilouise Kroker, 'Theses on the Disappearing Body in the Hyper-Modern Condition', in Arthur Kroker and Marilouise Kroker (eds), *Body Invaders: Panic Sex in America*, New York: St. Martin's Press, 1987, p.31.

6 Ibid., p.22.

7 See Katherine N. Hayles, *How We Became Posthuman: Virtual Bodies in Cybernetics, Literature, and Informatics*, Chicago, IL: University of Chicago Press, 1999.

Hayles suggests that Western culture has been determined within these relations of presence and absence, and the experience of living in Western culture. Bodies were there (present) and not there (absent). These relations between presence and absence meant that there was some stability to social order. Now this has all changed, this is not to say that presence and absence have lost all their meaning. Nor is it to *ignore* that the patterns and randomness *are* crucial to radical change, in their prominence within culture at large and to a radical change in the way that we conceptualise the world, our bodies, and our machines.

Information-based technologies are based precisely upon the blurring of presence and absence and a mixing of them into patterns and randomness. In cyberspace, it is the pattern that one can recognise and manipulate that is more important than one's actual presence or absence. Information, then, even if materially 'in' the world, is something conceptually distinct from its carrier. To the information carrier (the body), presence and absence may matter a lot. To information itself, however, it is the pattern formed that matters. Hayles suggests that information, the basis of so much in contemporary society, is never really present in itself. It is rather something we get out of the patterns formed.[9]

Varela, Thompson and Rosch, writing in *The Embodied Mind*, suggest that 'embodied cognition' occurs through 'inaction'. For them, cognition 'depends upon the kinds of experience that comes from having a body with various sensorimotor capacities' which are in turn 'embedded in a more encompassing biological, psychological, and cultural context'.[10] Because all are so mutually influencing, there is no easy separation between carriers of information and information between perceptual systems and perceptions, between agencies of thought and thought itself. It is suggested that these are all produced in tandem, through the various connections between bodies in what is called 'structural coupling'.[11]

---

8 K. Hayles, 'Flesh and Metal: Reconfiguring the Mindbody in Virtual Environments', *Configurations*, 10(2) (Spring 2002): 208.
9 Andrew Murphie and John Potts, *Culture and Technology*, Basingstoke: Palgrave, 2003, p.126.
10 Francisco J. Varela, Evan Thompson and Eleanor Rosch, *The Embodied Mind: Cognitive Science and Human Experience*, Cambridge, MA: MIT Press, 1991, p.173.
11 Ibid., p.151.

Here mind is seen as an 'emergent and autonomous network' or 'system', 'constrained by a history of coupling with an appropriate world'. Anything perceived is only done so through perceptually guided action. The structures that support thought, including the basic biological structures (such as certain patterns of colour reception which vary from species to species) emerge from such action over time. In other words, perception has emerged from engagement with the world, and in turn, this perception seems to give us the world as we know it. Thought is not something to be produced in a laboratory, separate from the world. The brain, body and world all work together.

Although the perception that cyberspace is disembodied is refuted by the material realities of the situation, it nevertheless has a material effect on what technologies will be developed, how they will be used, and what kind of virtual worlds they will instantiate. As Sherry Turkle puts it, the culture of AI research is 'deeply committed to a view that thought does not need a unitary agent who thinks'. It is in this sense that Turkle suggests 'the computer takes up where psychoanalysis leaves off', taking 'the idea of a decentered self' and making it 'more concrete by modelling mind as a multiprocessing machine'.[12] However, this critique is only accomplished through the transcendence of not only bodily limits but the body as such, as if the body could not be thought outside a concept of organic unity as eulogised by Deleuze and Guattari's attack on 'the organic organization of the organs'.[13] In current discourse, even the body itself would no longer seem necessary; rather what must be recognised is the insistence on 'organs instead of bodies' (OiB) – namely, organs that are configured as 'inside out', having lost their quality of being 'in' the body.[14] What counts foremost in current analysis is that this OiB is a 'flattened' body that has attained the value of a screen, a surface of reflection – in other words, a medium in itself: the medium has taken the place of the body.

---

[12] Sherry Turkle, *The Second Self: Computers and the Human Spirit*, New York: Simon and Schuster, 1984, p.309.

[13] Gilles Deleuze and Félix Guattari, *A Thousand Plateaus: Capitalism and Schizophrenia*, London: Athlone Press, 1988, p.158.

[14] Bernadette Wegenstein, 'Getting Under the Skin, or, How Faces Have Become Obsolete', *Configurations*, 10 (2) (Spring 2002): 223.

A series of related questions emerge. Do the technologies of posthumanism participate in a narrative of disembodiment or one of 'refigured embodiment'?[15] In the speculative discourse of cyberspace, we are promised whatever body we want, which doesn't say anything about the body that I already have and the economy of meanings I already embody. Generally, if we look to those who are already participating in body reconstruction programmes – for instance, cosmetic surgery and bodybuilding – we find that their reconstructed bodies display very traditional gender and race markers of beauty, strength and sexuality. There is plenty of evidence to suggest that a reconstructed body does not guarantee a reconstructed cultural identity. Nor, as Anne Balsamo claims, does 'freedom from a body' imply that people will exercise the 'freedom to be' any other kind of body than the one they already enjoy or desire.

Popular culture has eagerly embraced the idea of artificial sexuality as seen in various films such as *Brain Storm, Lawnmower Man, eXsitenZ* and *The Matrix*. The pleasure of the interface leads us into a microelectronic imaginary where our bodies are obliterated and consciousness is integrated into the matrix maternal-womb leaving our meat bodies behind. Magazines such as *Future Sex, Mondo 2000, Wired, Cyber-Edge Electric* and countless others examine how technology can facilitate cyber-sexuality... sometimes featuring high-tech sex toys in photographs, drawings, essays and stories. This 'remote control masturbation' has become a marketing gimmick, lending an aura of novelty to otherwise conventional products.

Predictably enough, it is when new technologies are popularly represented as having sexual consequences that the contradictions surrounding them and their cultural implications become most clearly visible. Rather than new relations between men and women, Net ethics throws into question

---

15 In *Volatile Bodies: Towards a Corporeal Feminism*, Bloomington, IN: Indiana University Press, 1994, p.5-13, Elizabeth Grosz offers an extended reading of this Cartesian tradition in contemporary philosophy as she attempts to define a feminist alternative to it, one that does not depend upon some version of the mind–body dualism. See also Allucquere Roseanne (Sandy) Stone's comment, in 'Split Subjects, Not Atoms; or How I Fell in Love with my Prosthesis', *Configurations*, 2 (9) (1994), that 'the physical/virtual distinction is *not* a mind/body distinction'. Instead, she argues that the virtual is 'a different way of conceptualising a *relationship* to the human (or, for that matter, the transhuman or posthuman) body'.

essentialised identities and dualistic sexual categories. Notions of authenticity, of essential femininity and of the self are displaced in favour of Baudrillardian multiple roles, alternative personae and a matrix of potentialities which allows people to recode themselves ahead of disciplinary technologies.

Erotic interfacing, is after all, purely mental and non-physical; it is always a free play of imagination such as in MUDs or MOOS, or chat lines such as Sappho. Gender swapping or virtual cross-dressing in MUDs is quite common. Any account of the evocative nature of gender-swapping might well defer to Shakespeare who used it as a plot device for reframing personal and political choices. Deconstructing femininity is a game that female MUDers play. One player, a woman currently in treatment for anorexia, describes her body this way:

In real life, the control is the thing. I know that it is very scary for me to be a woman. I like making my body disappear. In real life that is. On MUDs, too. On the MUD, I'm sort of a woman, but I'm not someone you would want to see sexually. My MUD description is a combination of smokes and angles. I like that phrase 'sort of a woman.' I guess that's what I want to be in real life too.[16]

We might view this aestheticisation as precipitating a 'crisis' of identity or perhaps, more accurately, as the source of an ongoing tension or discomfort in the vulnerability towards health and the gendered-race marked body. But this induction of the surfaces of the body into a state of enforced signification seems to have been more novel for males, given that women's bodies have had a longer history of subjection to inscription of idealised images. This sense that the female body has long been constituted through a dense accretion of simulation model informs Kroker's deliberately provocative assertion that 'women's bodies have always been postmodern'.

---

16 S. Turkle, *Life on the Screen: Identity in the Age of the Internet*, New York: Touchstone Books, 1997, p.215.

Yet, the sexism of the relations between women's bodies and the production of the new technologies is often belied by the slick presentation of aestheticised images. Though I oppose Net censorship, I don't think it's unfair to suggest that this type of print exploitation leaves women vulnerable to the propagation of 'virtual sex' and other sex-peddling services offered on the Internet. Despite the rhetoric that claims such games are 'harmless', they reify a cultural way of looking at women, which is destructive and demeaning. Studies show that the anonymity offered by the computer screen empowers antisocial behaviours such as 'flaming' and borderline illegal activities such as trespassing, e-mail snooping and MUD-rape.

Consider another virtual myth: sexism against women may be reduced through CD-ROM and mediated reality and augmented reality programs because men can now log-in as the female and 'take' the woman's point of view. This speculation sounds encouraging, but it essentially ignores the fact that identities of sexuality are constructed socially in ways that cultures powerfully inscribe bodies. Women are socialised in a manner that can't be replicated by assuming 'different point of view'. Nor can the sexual hierarchy be overturned so simplistically, for Western culture continues to privilege the male within the confines of the phallocentric tradition.

Though early representations of cyberspace denied sexual difference by positing the Net as a 'gender-neutral zone', scholarship has focused on issues specific to gender differences within various postmodern technologies. Some speculate that, like technology, cyberspace itself – what I think of as an externalisation of symbolic code – is masculine.[17] Margaret Morse describe cyberspace as hermaphoroditic, 'divided by gender inside and outside. The interiority of cyberspace, like the interior of a cave, is like being enclosed inside the womb.'[18] Furthermore, the interfaces of cybernetic space have been

---

17 Andreas Huyssen explained why mass culture, technology itself and machines are gendered female in his *After the Great Divide: Modernism, Mass Culture, Postmodernism*, Bloomington, IN: Indiana University Press, 1986. For a male cyberspace, see Rob Milthrop's 'Fascination, Masculinity and Cyberspace', in *Immer-sed in Technology*, p.129-50. Gillian Skirrow's 'Helli-vision' draws on Klein to describe a male and female relation to the game world as inside the mother's body.
18 Margaret Morse, 'Virtually Female: Body and Code', in David Trend (ed.), *Reading Digital Culture*, Oxford: Blackwell Publishers Ltd, 2001, p.90.

imagined as a seductive and dangerous garment. The fantasy of putting on such a second, virtual skin is said to express a longing 'to become woman'.[19]

Claudia Springer discusses the 'feminisation' of virtual technologies, which she distinguishes from the 'masculinisation' of industrial machines and technologies. Specifically, she relates the invisible networking of cyberspatial systems to 'conventional ways of thinking about female autonomy and feminine subjectivity'.[20] Drawing on Thomas Laquer's two-sexed model of human sexuality, Springer deconstructs women's inferior sexual status by arguing that women's bodies are read as the opposite of men's, due to their lack of male genitalia, as well as the association of women with inferior, fluctuating spaces of the body. Springer sees the feminisation of virtual technologies as central to understanding issues of identity and embodiment within cyberspace itself.[21] Interestingly enough, we see human actors replaced by their digital counterparts. In 2000, telecommunications giant Motorola commissioned Digital Domain to create an animated personality as the face of its new voice-activated web browser. The character of Mya is based on actress Michelle Holgate, who was extensively photographed and the images scanned into the computer. Amazingly, the initial results were so lifelike that viewers were unimpressed because they thought it was a real woman. So a revised version was made to look more artificial, with shiny skin actually modelled on the lustre of a china plate.[22] With respect to virtuality, it no longer makes sense to ask whose reality/perspective is represented in the various simulated worlds, the industry, or the subculture; rather, we should ask what reality is *created* therein, and how this reality *articulates relationships* between technologies, bodies and cultural narratives.

The play with reality and virtuality, the deconstruction of the concepts of interiority and exteriority as presented in the

---

19 A. S. Stone, 'Will the Real Body Please Stand Up? Boundary Stories about Virtual Cultures' in M. Benedikt (ed.), *First Steps*, Cambridge, MA: MIT Press, 1991.
20 Claudia Springer, 'Muscular Circuitry: The Invincible Armoured Cyborg in Cinema', *Genders*, 18 (1993): 92.
21 Quoted and discussed further in Dawn Dietrich, '(Re)fashioning the Techno-Erotic Woman: Gender and Textuality in the Cybercultural Matrix', in *Virtual Culture: Identity & Communication in Cybersociety*, London: Sage Publications, 1997, p.170.
22 Cited in Marie O'Mahony, *Cyborg; The Man-Machine*, London: Thames & Hudson, 2002, p.68.

works of Diller + Scofidio (*withdrawing Room*, 1986), Aziz + Cucher (*Interiors* 1999), Brandon Stosuy (*Cam0005*) and Motorola among others, brings an important dimension into the picture – namely the desire to blur the digital with the real, leading some such as Kevin Warwick, Steve Mann and Orlan to get involved with their own skin. Orlan invokes a psychoanalytical interpretation of the skin as a platform between the image of the self and the body that one is. She summarises this representational dilemma in the sentence 'I am never what I have.'[23] But she also invokes another important issue – namely, the impossibility of becoming one with one's proper image in the mirror or on the screen. For Orlan, it becomes necessary to masquerade herself, putting on the 'skin' of different icons throughout the centuries, just as Alba d'Urbano hung her own 'skin-suit' on a coat hanger. What Orlan, as Wegenstein notes, is pointing to in her performance art is that 'behind the masquerade there is nothing, a "no-space"'.[24]

This dilemma of 'what-lies-behind-it' has also been thematised before 'cyberspace' and the 'virtual body' by the Austrian media artist VALIE EXPORT in her film *Syntagma* (1983). What is alluded to here is a cinematic space that by definition remains empty. This space resembles the 'empty image' that Marie-Luise Angerer speaks of in her book *body options*: an image that remains empty because it is 'behind' the body image – it is a surplus to the image, and reveals itself only through its absence, the absence of the signified. This emptiness or 'no-space' creates an unreal subject position, for it cannot be looked at from any possible position in real space. As Angerer puts it in her analysis of Orlan's performance art, this is a representation that never takes effect in an unbroken way.[25] Orlan emphasises precisely the empty position created by these 'breaks' in the symbolic order. These have to be masqueraded constantly – and most prominently – by the event of gender formation.

The virtual body is neither simply a surface upon which are written the dominant narratives of Western culture,

---

23 B. Wegenstein, 'Getting Under the Skin', p.248. Orlan is alluding to Lacan's distinction between having and being a phallus.

24 Ibid., p.248.

25 Marie-Luise Angerer, *body options: körper.spuren. medien.bilder*, Vienna: Turia and Kant, 2000, p.91.

nor a representation of the cultural ideal of beauty or of sexual desire. It has been transformed into the very medium of cultural expression itself, manipulated, digitalised, and technologically constructed in virtual environments. As the virtual body is deployed as a medium of information and of encryption, the structural integrity of the material body as a bounded physical object is technologically deconstructed. If we think of the body not as a product, but rather as a process – and embodiment as an effect – we can begin to ask questions about how the body is staged differently in different realities. Taken to an extreme, the awareness of the mediated nature of perception that simulacra technologies provide can be taken to signify that the body itself is prosthesis.

In short, what these virtual encounters really provide is an illusion of control over reality, nature, and especially over the unruly, gender- and race-marked, essentially mortal body. It is not a coincidence that virtual reality emerged in the 1980s, during a decade when the body was understood to be increasingly vulnerable (literally as well as discursively) to infection as well as to gender, race and ethnicity. With virtual reality we are offered the vision of a body-free universe. Despite the rhetorical disclaimers that this was not a Nintendo war, media coverage of the Persian Gulf spectacle provided numerous examples of the deployment of a disembodied gaze to mask the violence of reality.[26] Even if virtual technology promises a new form of subjectivity, it contributes to a heretofore unknown epidemic of cultural autism. Intimacy is now refined as a quality of interaction between the human body and the machine.[27]

---

[26] The '24 Feburary 1990 Doonesbury strip' by Gary Trudeau offered a frame-by-frame depiction of the bomb's-eye view of a bomb travelling into a chemical weapons facility 'past startled Iraqi production managers and into the office of the facility administrator'. The next frame indicates an explosion, while the narrator (a general in the next frame) states: 'Unfortunately, it continues through an open window and explodes in a nearby parking lot.' Earnest Larsen considers the implications of what we didn't see during the television coverage of the Gulf War. Ernest Larsen, 'Gulf War TV', *Jump Cut*, 36 (1991): 3-10. Today, after the second Gulf War, the violence erupted on the screens and made itself impossible to mask.

[27] Sherry Turkle and Seymour Papert argue that computer technologies may promote the development of epistemological pluralism. The most optimistic prophecy about virtual reality technologies would be consistent with their argument. But they go on to remind readers that the computer culture may inhibit the realisation of such possibilities. Sherry Turkle and Seymour Papert, 'Epistemological Pluralism: Styles and Voices within the Computer Culture', *SIGNS*, 16 (1) (Autumn 1990): 128-57.

If 'the frontier' functions metaphorically to describe the space of the disembodied 'social' in a hypertechnological informational society, cyberspace offers a way to think about the location of the social in post-industrial capitalism. Although this space is structured, it is impossible to map; there is no Archimedian point from which to construct a totalising vision of the scene. At best you can wander through it, reading/writing as you walk, and maybe stumble upon something that was not programmed for you.

Patriarchal representations of femininity, whether through Internet communications, Cyberpunk literature (although this is slowly changing) or CD-ROM games, are notoriously sexist; and yet these text and graphic-based images are occupying the virtual spaces where women could be defining their own relationships to the new information technologies. Such lack of representation carries with it significant consequences, not the least of which involves the shift of political power from an industrial basis to one in which knowledge or the communication of information operates as the new 'capital'. The exclusion of women from the technologies which produce and regulate this knowledge disempowers them in fundamental ways, especially given that virtual technologies are controlled, and managed on several powerful fronts: the government (especially defence), the corporate Western front and the media.[28]

While acknowledging the admirable accomplishments of female corporate managers, software developers, media artists and scholars, it is important to recognise that women in the larger culture are still marginalised at the periphery of the communications industry with unequal pay. The social biases established early in education, which separate women from the disciplines of maths and science, represent deeply embedded cultural prejudices against women in technical sciences. There are, of course, numerous organisations that promote female-oriented domains in cyberspace and practioners in digital media. Their work often involves considerable personal sacrifice and seems to be sustained by a belief

---

[28] Cited in D. Dietrich, '(Re)fashioning the Techno-Erotic Woman', p.176.

that the redeeming power of artistic expression should be made available to those who are excluded from the mainstream of information society. Their labours have a limited visibility, since they are collaborative and especially when there is an ongoing institutional relationship to support them. The common denominator of all these women is that they did not occupy a niche or serve a pre-existing group, but crossed cultural boundaries and oppositions to create new domains that include technological have-nots. Their work has a feminine flavour – a cave building and garment fashioning that envelops or wraps a domain of technology and artistic creation.

Not surprisingly, the notion of cyberspace as a radical domain for women raises several issues of concern. As some cyberfeminists have argued, female bodies are inscribed culturally into specific paradigms that determine the nature of identity and subjectivity. For women in Western cultures, this has been a paradigm fraught with difficulties, for the physical body has been the site of fervent battles regarding female sexuality, reproduction and identity, so much so that it becomes difficult to separate feminine subjectivity from a particular system of embedded power relations. Because of these patriarchal tensions, it is necessary to situate female subjectivity within a gendered and politicised context in the cybercultural matrix. Put differently, women stand to gain little as quasi-disembodied subjects within a network environment *without reference to the material conditions of their subjectivity*.[29]

Yet, projecting a gendered self into cyberspace can be a very painful experience for women. Issues involving gender politics and representation cut to the core, and many involving women are simply tired of buttressing the same fronts time and time again. In short, there do not seem to be easy answers to these difficult social issues. Drawing on Donna Haraway's mythos of the cyborg, I imagine an electronic space that is about 'transgressed boundaries, potent fusions, and dangerous possibilities', called for in order to resist dominant (rational) power structures embedded within patriarchal culture.[30] In

---

[29] Ibid., p.178.

[30] Donna Haraway, *Simians, Cyborgs and Women: The Reinivention of Nature*, New York: Routledge, 1991, p.154.

this instance the fusion of machine and organism becomes a progressive, and transgressive, hybrid – an artificial site for ongoing political activity, which necessarily involves deconstructing 'incompatible' frames of reference. Haraway powerfully identifies the corpus of the cyborg as a self-conscious coalition – a political kinship forged from radical feminist initiative and action – and necessarily involving difference and contradiction. For Haraway, building effective unity doesn't eradicate the revolutionary subject, for the permeability of boundaries in both the body and the body politic assures transgressive leakages as well as radical fusions. New couplings must, of necessity, bring about new coalitions.

The question then becomes how to ally the new technologies with progressive political movements. And here, Haraway, Sadie Plant, Kathy Acker among others point to the empowerment of feminist textuality. What is particularly interesting about Haraway's conception is that such political empowerment is constituted from textuality – in other words, from women's collected voices, stories and myths. And it is here that I believe a community emerges within the cyberspatial matrix: women of different ethnicities and backgrounds (depending on access) join the circuitry of the electronic network, responding to one another's dialogue through digitised conversation.

In her book, *The War of Desire and Technology at the Close of the Mechanical Age*, Stone acknowledges the relationship of physical bodies to the epistemic structures by which they become encoded in culture, and she is particularly careful to ascribe gender discourse and meaning to the physical body itself, an embedded cultural phenomenon. She identifies this concept as 'a body unit grounded in a self' and suggests that telling any personal narrative seems to depend upon this material identity.[31] In other words, situated between a

---

31 A. S. Stone, *The War between Desire and Technology at the Close of the Mechanical Age*, Cambridge, MA: MIT Press, 1995, p.84-5.

Right: Poster for 'How to get out of one's bedroom', a self-support group, initiated in Paris, by ALD (A Longue Durée—A Long Lasting Project) which proposes an insight into the everyday life of female homelesses and squatters. (Credit: ALD)

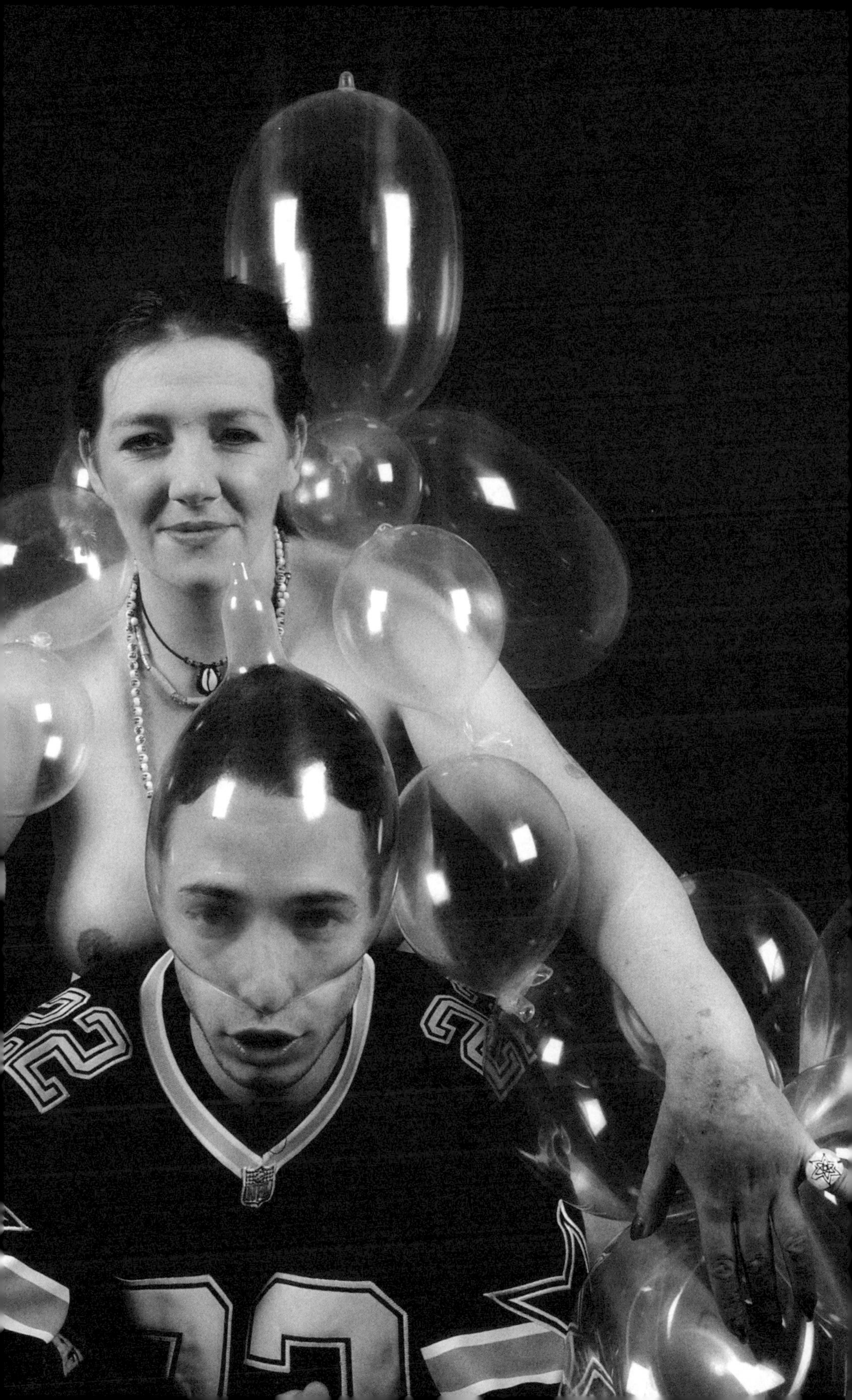

gendered, material body and an ethereal, cyberspatial identity; between patriarchal culture and feminist community; between inside and outside, the female participant must embrace ambiguity and conflict in order to appropriate a cultural space for feminist discourse. Because the space is electronic, such a tactic necessarily involves issues of subjectivity and embodiment. Feminist theorists of technoscience have argued that technologies are never neutral and apolitical. Rather, they are structured by their economic, political, and social contexts, and they are 'haunted', in Patricia Clough's terms, by our histories, languages, memories and unconscious desires.[32] Further, technologies create privileges and constraints, and access to and control of technology are highly political matters.

Given the need for continued development of feminist communities amidst dwindling resources, cyborg politics demand that we reimagine social and political possibilities for communicating through electronic media – that we utilise and mobilise the powerful venue that cyberspace offers. This means working within existing structures, such as the cyberspatial print industry, its electronic counterparts, educational, governmental and corporate institutions. It also means continuing to cultivate the margins of digital culture, where the greater experimentation is taking place: for in this post-industrial present we are left with radical information technologies, possibilities for new social matrices, and issues of textuality and gender to explore further. Though we stand to gain little by idealising these new technologies, we can embrace their differences in the hope that they might introduce new ways to textualise both our social spaces and our bodies, allowing us to reimagine a feminist politics of the future.[33]

[32] Patricia Clough, *The End(s) of Ethnography: From Realism to Social Criticism*, 2nd edn, New York: Peter Lang, 1998, p.xxv. Cited in Victoria Pitts, In the Flesh: *The Cultural Politics of Body Modification*, Basingstoke: Palgrave, 2003, p.176.

[33] Cited and discussed further in D. Dietrich, '(Re)fashioning the Techno-Erotic Woman', p.182.

Chapter 16

Francesca Hughes

Hugh Lawson-Tancred

STABAT MATE
on standing in for matter

Aristotle

Chapter 16 — Francesca Hughes

ER:

John Milton
Catherine Ingraham

Tom Phillips; Jennifer Bloomer

Steven Johnson

Bertrand Russell

STABAT MATER: on standing in for matter

Preface

This chapter was first given at the *Alterities* conference in the Paris Summer of 1999. I can calculate the date by the age of my second child, with whom I was just pregnant. The conference came at (what I was later to learn was) a climax in architectural culture of work and publications and other conferences, a veritable watershed of both production and thought, on the intersection of gender and architecture. When, some three years later, I emerged from my *sejour* in early motherhood, I discovered to my astonishment that the world of architectural culture was possibly even more changed than I: all trace of the ten years or so of work that had preceded, culminated almost,

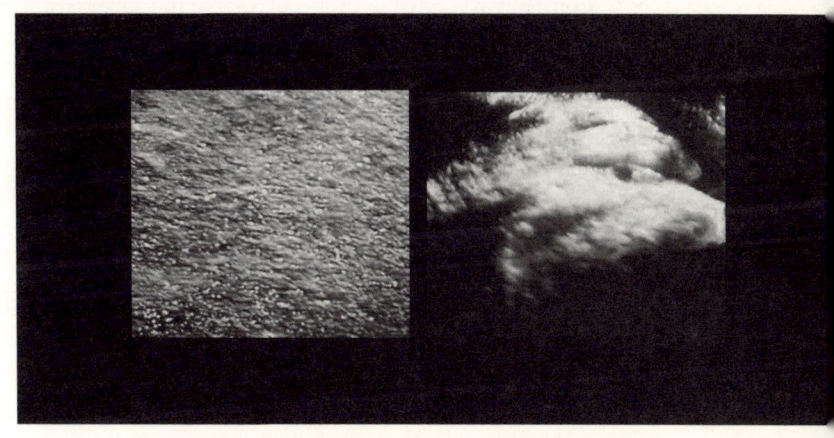

(The following sequence of images pp.261-279 are courtesy of BBC Worldwide)

in the *Alterities* conference, had disappeared. 'It's as if it never happened', a colleague offered in ambiguous comfort.

It was Bloomer who, with her penchant for what Americans call 'dirt', brought matter to the table. In the work it inspired in many, including myself, this great, muddy, parting gift generated footprints in various directions. With hindsight, one can now see that this was one of a handful of beginnings that laid the ground for not only a certain importation of Bataille's *L'informe* that was to follow, but also for the current interest in emergence or self-organising systems with its inherent sympathy for the lowest common denominator. The most basic component, after all, at the very bottom of 'bottom up' intelligence, is matter. Nothing is lower. Bloomer, and before her, of course, Mary Shelley, reminds us that form's superiority over matter, especially in the moment of production, is not necessarily as stable as it might appear.[1] That matter might harbour desires and intelligence of its own. This undoubtedly feminist legacy is now embedded without trace in the discourses on emergent systems and self-organising structures that are the currency of the moment in the engines of architectural culture. Though the work on feminist critical theory and architecture that took place in the 1990s has somehow been sublimated into the body of architectural thought, we can track its crucial legacy when we look at how contemporary discourses are inflected by, among other things, a sympathy for matter. Consequently, one might speculate that the papers from the *Alterities* conference are potentially of more import and interest now, after this transforming interval, than they were immediately after the conference. 'Time', to quote Tom Phillips' painting of the same title, 'is a great dealer'.[2]

For my small part, this chapter is a too densely packed *valise* of half-formed thoughts that have been unfurling in my own work on 'error' these past few years. I have left the text pretty much unpacked, as a marker of what was going on before everything fell quiet.

At about the time of the first ultrasound (I was sixteen weeks into my pregnancy) I found myself surveying yet again a very old but rather banal building in France (the site of a project for an artist's studio). The true location of the fabric that makes up this building had successfully confounded me on two previous occasions. Apparently plumb walls when drawn up would sway and twist, plans flared over each other, sections slid around... No amount of precision, thoroughness or patience, it seemed, could fix this house to the page. I began to eye the matter that made up these walls with some caution, and not without a touch of paranoia. These seemingly banal

---

1   In returning to *Frankenstein; or, The Modern Prometheus* with matter in mind, it is breathtaking how neatly Shelley elaborates a poetics, an evidence almost, of an intelligence that originates in matter itself. Quoting from *Paradise Lost* in the opening pages, she reminds us of Adam's reproach:

Did I request thee, Maker, from my clay
To mould me man, did I solicit thee
From darkness to promote me...?

Implicit is the possibility that matter *could* actually solicit form, corrupt form and then cause its downfall, the very scenario that Shelley unfurls, to Victor's own horror. John Milton, *Paradise Lost*, Book X, p.743-5.

2   Tom Phillips, *Time is a great dealer*, 1988.

rooms were encased, floated, in a crust of extraordinary stuff whose form proved bizarrely unreadable, or at least unwritable. What was one to do with the acute fear that form itself was absent and one was dealing directly with matter? Absurdly trying to measure, plot and calculate formless[3] matter. Speaking the wrong language to it, and getting at best no answer, and at worse, a different answer each time.

I have a printout, a still, from the ultrasound. Constellations of white specks make up the lining of an interior. Within this two clouds of specks have drifted away and settled next to each other. That is all that they are; they are also, unmistakably, two perfectly formed, crossed, tiny

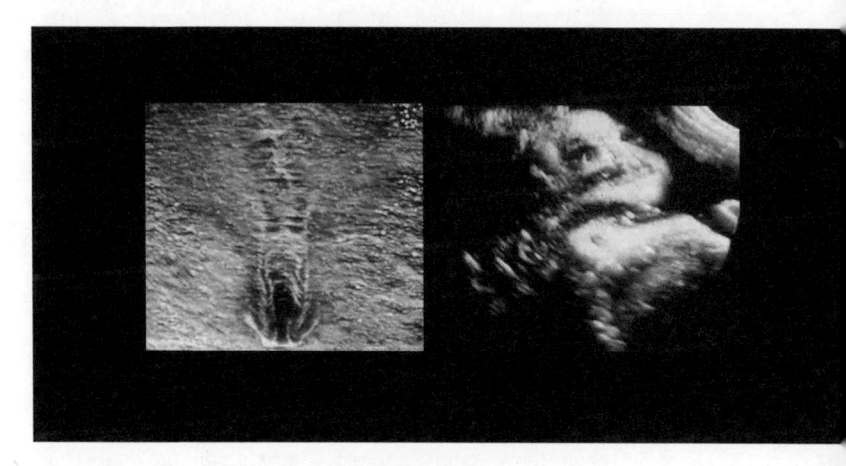

feet. Almost classical in their poise. There is little of Plato's cavernous interior to this luminous, red, noisy, swaying, pulsing enclosure. As the microphone rests on my jelly-coated belly the apparent walls of a space within collapse and reconfigure, cleaving continuously, a further wall, or is it a body within a void, twists and turns, to reveal its own voids. In trying desperately to make sense of this moving picture, solid/void, exterior/interior, form/matter become such redundant registers that one is again aware of addressing a problem in what is simply the wrong language. The constant

---

3 Since the return of Bataille's *L'informe,* the term *formless* has lost the innocence wanted here. More relevant perhaps is the 'formless morass' the Materialists (contemporaries to and opponents of Plato) referred to in attempting to describe the physical universe beyond our 'form lending' senses. See Hugh Lawson-Tancred's Introduction to his translation of Aristotle's *Metaphysics,* London: Penguin, 1998.

state of cleaving relations that the ultrasound depicts is unnerving. It shakes me to the core. How can one shrug off the sneaking suspicion that all along we've been doing it not quite wrong, but not quite right either. That our methodologies glance off, without ever really penetrating, what it is we are looking at. That they get us *kind of* there, but not exactly where we need to be.

The overt presence of form (and all things formal) in architectural discourse and culture is marked by the absence of matter. It is an absence so complete that the silence, once heard, roars and cannot be shut out. So subsumed by form is matter that in the form/function paradigm we find matter has all but disappeared, although we know it's in there, lurking within the term 'form'. But if we prise form apart to call matter to the fore, we find it reluctant, it will not come.

Instead, matter enters architecture, repressed, and consequently fetishised, as materiality;[4] and illegally, as error. These two modes of access are not unrelated (as the intersection between materiality and error demonstrates) but play very different roles in the complex architecture of matter's containment. Of course, fetishisation aside, materiality provides architectural production with a carefully constructed vehicle able to carry matter, all the while preventing direct contact (read contamination): matter in materiality is already at a remove, safely isolated. There is a degree of abstraction (i.e. form) at work. Innate and extraordinary properties – viscosity, ductility, brittleness, surface tension, expansion rate and freezing point – are contained, as a threat might be, by coefficients. A pale-like structure surrounds the unruly matter. If matter is, in Aristotelian terms, potential form, it might become form and it might not, materiality presents matter irreversibly poised to become form: in the wings, no turning back, at our disposal. The doubt or threat that promised form might be lost and matter return is removed. Materiality is matter domesticated, successfully colonised by

---

[4] This fetishisation is typically organised by desire (here nostalgic) and absence (only imagined) as Jennifer Bloomer so cannily points out: 'In its subjugation of matter by form, the modern concept of design is necessarily dominated by a nostalgia for matter, a fetishisation of an imagined absence.' See her wonderful essay entitled, 'The Matter of the Cutting Edge', in K. Ruedi, D. McCorquodale and S. Wigglesworth (eds), *Desiring Practices*, London: Black Dog Press, 1996, p.18.

formal criteria. If matter is birds in the bush, materiality is birds firmly in the hand. We well know which is worth more.

Materiality then acts as a kind of lever on matter where the matrix of coefficients that materiality represents, and the complex architecture of expectations it projects around a specified material, is crucial. Performance calculations locate a boundary between correct and erroneous behaviour: a contract between form-making and the material (not matter) in question. But this pale of coefficients that is materiality is a structure under severe strain and consequently frequently breached: matter that exceeds the constraints of materiality (and enters the realm of 'material failure'), matter *beyond*

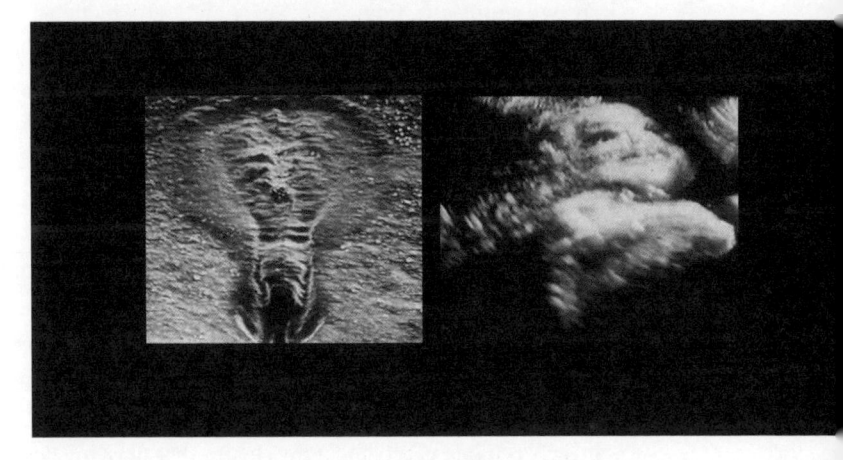

the pale that is, is error. (Or to be more precise, matter that is *identified* beyond the pale is error, as there is also presumably the matter that gets away.) The architect, however, has anticipated this and is fully prepared. Allowances have been made, even designed, the space of error is dimensioned and located in the drawing. The degree of tolerance appropriate to not only the materials and construction techniques involved, but also to the physical and economic site conditions, is selected and its allotted territory literally plotted into the working drawings. Here, like a safety net, the escaped matter is again contained. The margin for error, itself a palisade, insulates the calculated from the incalculable. Any breach of this confinement is neatly construed as error where error indicates a lapse identified, be it in methodology and technique, or in the predictability and uniformity of a material's properties, or in the extraordinary hermetic seal that wraps the project on and off site, isolating it from all that architecture must keep out. The space of error shadows all material constructions, including this one.[5] The complex apparatus architectural practice sets up to hold it at bay, paradoxically only serves to articulate it further: the very use of the term 'tolerance' in the first place signals intolerance, as there is only the need to tolerate something when it is deemed intolerable. Matter is.

It was very clear what needed to be done to this building (I tried to forget it was a house, that extraordinary cathexsis of everything that architecture complicates). It needed straightening out, tidying up. Space and light, hygiene, weightlessness, smooth seamless surfaces, convenient and efficient dispositions of parts, in short: firmness, commodity and delight. I had the whole canon in my bag and was ready to wreak revenge on these disobedient walls, and floors, and ceilings… I proposed a vertical slot (space and light) to release us from the domestic lid-like ceilings, a sanitary unit (hygiene), suspended between floors of course (weightless, let's be done with this gravity nonsense), wrapped in a stressed steel skin (smooth and seamless)… and all in hot orange for delight… I took my drawings back to London and with 1500 km of conveniently abstracting distance, and the utterly reliable precision of my clean white drawing board, my 0.18 pen, my perfect eyesight, performed architecture on this unwieldy body. At the same time, I started to feel a somersaulting in my belly, the turnings of solids in voids in solids in voids, unrepeatable choreographies in unimaginable (or perhaps only imaginable) spaces. I glimpsed the utter precariousness of my architectural actions and reached for a finer pen.[6]

How can we explain the obsession in architecture with not precision per se, but with a certain aesthetic effect of the

---

5   The textual is of course not without the material, as anyone who has ever encountered the resistance text puts up against the process of editing or translation will know.

6   That the history of this chapter spans the final demise of drawing board and pen is not without irony. The economy of error and precision, so elaborated in the practice of architectural drawing, has not been rationalised by the shift to the digital but instead is curiously reincarnated, with a whole new set of behavioural possibilities for the category of error.

Stabat Mater: on standing in for matter

precise that exceeds the degree of precision necessary for the task at hand?[7] A precision that is frequently redundant (doesn't need to be there, functionally in terms of error reduction, or symbolically, in terms of holding up meaning), or fake (it isn't actually precise). Yet it is there, enduring, coveted, reiterated through the whole family of representations that surround the production of architecture. In such instances, this excessive precision can only be understood as standing in for something, but what? Architects are trained to mistrust matter, desist gravity and abhor error. In what is clearly a futile endeavour, we operate despite the knowledge that matter, and with it potential error, lie in wait. The architecture of the evasion that ensures itself structures almost all of our thinking about making buildings. We go to great lengths to both separate ourselves from and control the act of making buildings. These lengths, the production of complex documentation in order to direct construction by others, the legal acrobatics that allow delegation while still maintaining control, and lastly the discourses of architecture that patrol its relations to the built, after all, define the architect.

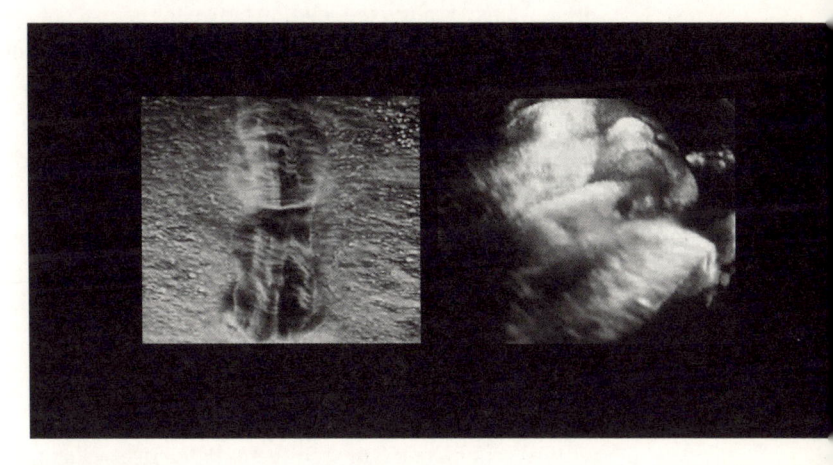

[7] At the level of the architect's construction drawing (production information), once the precision of the drawing meets the physical limit of what is possible on site, in the mud and rain, or in the manufacturing of a component, or at the junction between different materials and trades, every further degree of precision is, by definition, redundant: any further thinness of line, accuracy of radii, decimal point to dimensions exists instead in a margin of excess. A margin made all the more convoluted in its configuration by the recent development and increasing application of CAM-CAD production where the quantum gap between digital and material collapses at the interface of the two. Quite what this yields is still very unresolved and indeed untested.

The architect then does not make the building, but draws it. It is the drawing that makes architecture's separation from building bearable, and the drawing that carries the burden of this anxiety or loss.[8] And predictably, it is here, in the drawing, that any excess of precision finds its most elaborated manifestation. Any elaboration of the drawing is an elaboration of the site of the loss itself. Moreover, when we draw a plan or section we are reminded that both projections are understood as cuts. Both *want* to be cuts. When we draw a section, or plan, we are cutting through stuff, all kinds of stuff. Not only through certain conceits about precision and construction, through the relations between two and three dimensions, through the relation between the concept and the plan figure, the concept and the sectional figure, through the complex relations between economy and resource, manufacturing and commodity, through the relations between the engineer and the architect, through the legacy of cutting that architecture and medicine have traded back and forth throughout history, through the architect's desire for the building to be a seamless, uninterrupted extension of the drawing, but also through the desire to cut and bite and hold and engage with the material world that we were born into. To feel the resistance of lost matter meeting matter.

There is a context: the excess of *apparent* precision within certain aspects of architectural practice occurs within a culture in which the desire for the 'precise' is already dominant. We know already that whenever a tendency is excessive, it is likely to be signalling repression. The obsession with an 'effect' of precision that far exceeds the actual precision necessary to the making of buildings signals, like the first 'no!', the first moment of discipline, the fear of not working. 'Precision' is the key agent in the repression of matter by form. Like all forms of discipline, the less effective it is, the more excessively it is employed, and the more unbending, frequent and extreme its application is likely to be. Conversely, when discipline is extreme, it is often also at

8 For a dissection of a 'Rilkian' lament that ensues directly from this loss and pervades the discourses of architecture, see Catherine Ingraham's 'Losing it in Architecture' 'Losing it in Architecture', in Francesca Hughes (ed.), *The Architect: Reconstructing Her Practice*, Cambridge, MA: MIT Press, 1998.

a remove. The relations between the architect and matter itself are (by definition) remote.

Located at some depth, in the archeology that is the architectural drawing, is the boundary between effective precision and redundant precision. This latter precision that exists in a margin of excess, more akin to the abode of ornament perhaps, is of ambiguous function in that it isn't actually working to prevent the incidence of error and cannot be reduced to a pragmatic action, but instead 'functions' as a symptom of some internalised desire (or malaise) at some altogether other order. After all, effective or actual precision by definition cannot be excessive, because it must be present at *precisely* the

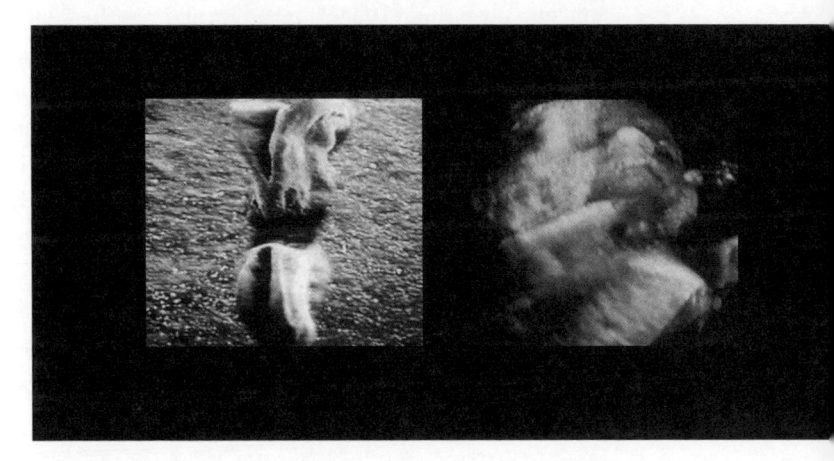

right degree. This mimicking of the precise that redundant precision constitutes then, has nothing to do with accuracy of production and everything to do with the organisation of the relations between form and matter. It is the external manifestation and aestheticisation of formal purpose uncompromised by reluctant, recalcitrant and unpredictable matter. The desire to mimic the precise, *and not to actually be precise*, is the desire to firmly repress the threat of matter directing, even for a moment, form. It is a mimicking which is so integral to the practice and culture of the architect, that we forget it is even there: we don't see it. It constitutes a form of control that is both about and because of relations that are remote. And the moment when these relations collapse is not only important but also catastrophic.

The gestating body, however, represents not the clean, efficient athletic body of Modernism, the body of the sanatorium and stadium, that emulates the precision of the machine it inhabits, drives, and operates, but, instead, matter out of control. That is, it represents not the aestheticised face of precision but what one might call 'precision thwarted'. Of course, if we were talking *actual* precision, the body in gestation is at its most precise; most breathtakingly efficient as the heart adjusts in size and power to meet the changing blood volume that increases by 50 per cent, most intelligently adaptive as the same heart changes its relation to gravity in order to lie on its side and make way for the rising stomach and intestines; most super-humanly athletic as the spine sways and the gait adjusts to accommodate a continually shifting centre of gravity and increase in load; most 'hygienic' as the antibody count soars and the placenta screens almost all pathogens. There is something important here for our thinking about these relations, a lucid moment that ends as does the pregnancy and, too quickly, the memory of pregnancy, but a doubt is firmly established that won't go away.

When in 1946 (of all years), Bertrand Russell optimistically published *History of Western Philosophy*, he pronounced, among other things, the end of matter as it had been known. Modern physics was here to stay.

Energy had to replace matter as what is permanent. But energy, unlike matter is not a refinement of the common sense notion of a 'thing'; it is merely a characteristic of physical processes. It might be fancifully identified with the Heraclitean Fire, but it is the burning, not what burns. 'What burns' has disappeared from modern physics.[9]

Instead of 'what burns', he goes on to say:

events, not particles, must be the 'stuff' of physics. What has been thought of as a particle *will have to be thought of* as a series of events. The series of events that replaces a particle has certain

---

9   Bertrand Russell, *History of Western Philosophy*, London: Unwin, 1979, p.65.

important physical properties, and therefore demands our attention; but it has no more substantiality than any other series of events that we might arbitrarily single out. Thus 'matter' is not part of the world, but merely a convenient way of collecting events into bundles.[10]

There is something both poignant and revealing about Russell's brave struggle to embrace this new kind of matter that we will now, 'have to' live with. This new matter, rendered immaterial by the course of modern physics, is so laden with nostalgia that he speaks bizarrely of events in terms of 'bundles'.[11] But matter has been immaterial before: right in the middle of Aristotle's *Metaphysics*, as an aside almost, an astonishingly prophetic moment of doubt is voiced:

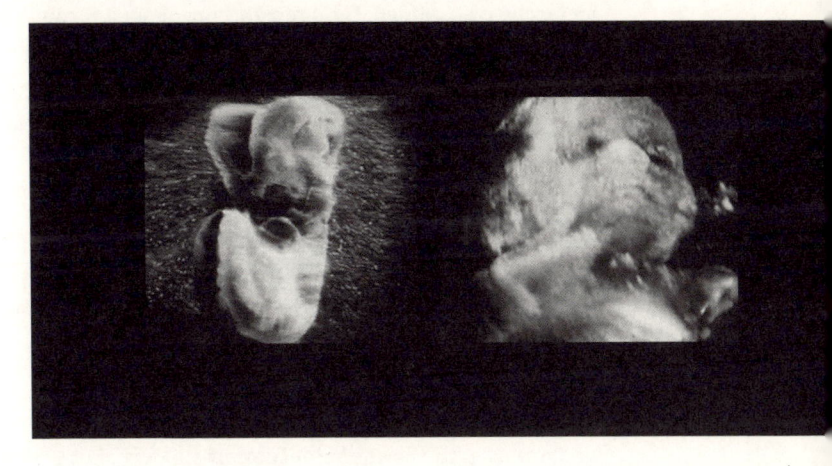

A question that can be deferred, though not for ever, is whether there is some other matter in addition to that of substances of the kinds that we have been examining, whether we should look for some other sort of substance, such as, perhaps, numbers.[12]

Now more than fifty years after Russell traded stuff for events, matter has re-entered physics with the curious advent of dark matter, that, like original matter, despite making up

10 Ibid., p.786, my emphasis.
11 A term that in the context of post-war imagery, of refugees carrying all their worldly possessions, is all the more laden with haptic sentiment.
12 Aristotle, *Metaphysics*, trans. Hugh Lawson-Tancred, London: Penguin Books, 1998, p.209.

99-odd per cent of the universe, we still can't lay our hands on. Rereading Aristotle, it is breathtaking again quite how little we have moved, how, like Alice, we have been running only to stay in the same spot. The matter that Russell so painfully left behind was the not so distant relative of Aristotle's matter, described first in *Physics* and later, more elaborately and more in terms of its relation to production (read form), in *Metaphysics*. The matter we inherit from Aristotle, and more importantly its relations to form stand: matter is potentiality, form is actuality. Actuality is prior (read superior) to potentiality, hence form subsumes matter.[13] In production, form and matter combine to produce a composite in which this hierarchy is sustained: form remains unchanged through the production, while matter is changed. Form pre-exists production: 'Form is not a product.'[14] Form is determinate, essential and above all teleological in a way that indeterminate, clay-footed matter is not.

So what *is* matter? Already matter is defined either by what it is not, or more curiously by what is undefined in the first place: what makes two men different when their form is both that of man? Matter. What makes man male and woman female? Matter. What makes one man black and another white? Matter. Matter is difference, is gender, is race, is everything that is difficult. And, matter itself is gendered: 'the man who brings in the Forms being himself single makes several things. And the male is in the same position in regard to the female. For the one is filled by many vessels, whereas the male fills many things.'[15]

Returning to Russell, the seamless purpose of this gendering becomes quickly evident:

> All change is what we should call 'evolution', in the sense that after the change the thing in question has more form than before. That which has more form is considered to be

---

[13] 'There can be no question that actuality is prior to potentiality' and 'the serious in actuality is both better and more worthy of reference than the serious in potentiality'. Ibid., p.272, 278.

[14] Ibid., p.199.

[15] Ibid., p.24–5. In Aristotelian terms, sperm is pure form and menses, the stuff of the matrix, is pure matter. Crucially, when Cicero translated '*hyle*' into Latin, he used the word '*materia*', which, like '*hyle*', means 'building material', and in doing so associated matter ever since with '*mater*' (mother).

Stabat Mater: on standing in for matter

more 'actual'. God is pure form and pure actuality: in Him, therefore, there can be no change... Only God consists of form without matter.[16]

Like the house that Aristotle repeatedly refers to as the exemplary composite of form and matter, the house of this construction is well built here: form is firmly in control of matter, or so it would seem:

Everything is produced either: (1.) From a bearer of the same name, as in the case of things produced naturally – an example of this among artifacts is a building which is produced from a

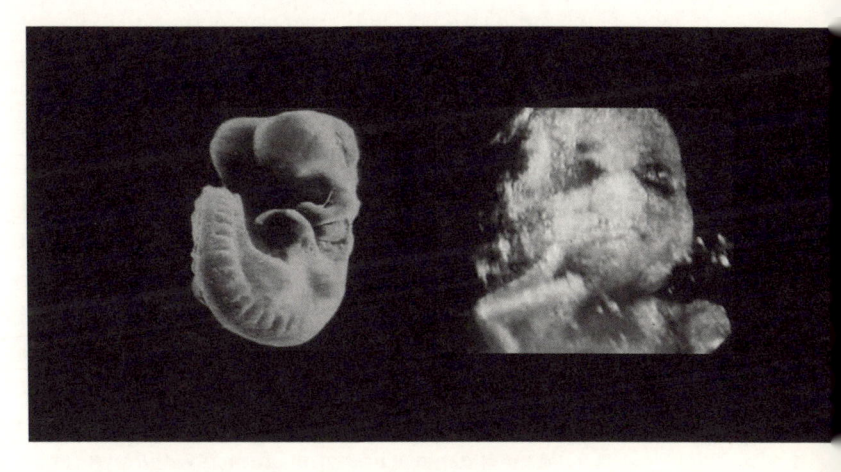

building to the extent that it is produced by thought, in that the skill is the form of the building – or: (2.) From a part with the same name or: (3.) From the possessor of some such part, ruling out, that is, mere cases of accidental production.[17]

'Mere cases of accidental production' – *this* is where the rain comes in, and matter leaks out. Like a craw that sticks in the throat of this treatise, this annoyance will not go away and resurfaces in various guises, as wine that turns to vinegar, as healthy bodies that become corpses, and most intriguingly, as accidental animals: the horse and the donkey that beget he mule (the sterile hybrid) that does not, because it cannot, beget the next mule. One senses these instances plague

---

[16] B. Russell, op. cit., p.181.   [17] Aristotle, op. cit., p.198.

Aristotle; try as he might, they won't bide the doctrine. All are instances where, unlike in 'buildings', matter makes mischief.

How is it that some things, such as health, are produced both by skill and by spontaneity, while others, such as buildings are not? The reason is that for some things the matter that provides a principle for production in the fashioning and producing of something that can be produced by skill, and in which an element of the item in question is present beforehand, is, in some instances, such as to be capable of being set in process by itself and, in others, not. And where it can be so self-initiated, it sometimes can, and sometimes cannot, be set in process in the way appropriate for a production. (There are, of course, lots of things that can be set in process by themselves, but not in some given way, such as, say, dancing.) [18]

The merry dance that matter leads Aristotle on is key here; it is the dance of accidental or self-initiated and above all spontaneous change.[19] Of the three categories of production listed in *Metaphysics* – natural, artificial and spontaneous – only the spontaneous resists closure and remains problematic. Here matter is asserting itself against form and the vector of change is firmly in the direction of degradation: rancid wine, rotting bodies, sterile hybrids. Eventually, perhaps in a moment of damage limitation, form is withdrawn, and matter left wholly to blame.[20]

When things go wrong, milk turns, the healthy body takes sick, wine becomes vinegar or the offspring proves sterile, there is, in Aristotelian terms, a privation of form, and matter rises to the fore. In other words, when things go wrong it is because of matter: the architecture of *Metaphysics* is the site of matter and error's intimate (and forced) alignment, a

18 Ibid., p.197-8.
19 What Aristotle is here distinguishing as 'self-initiated' change, 'things that can be set in process by themselves', apparently indeterminate or unscripted behaviour, is the signature characteristic of a system that is self-organised and that displays emergent behaviour: the movement from low-level rules to higher-level sophistication. (See Steven Johnson, *Emergence*, London: Penguin, 2001.) Implicit in Aristotle's distinction is the identification that in certain modes of production the architecture of form/matter relations is fundamentally different and, as in emergent behaviour, unpredictable.
20 Aristotle, op. cit., p.247: 'Any things, in fact, which in this way change into one another must do so by returning to their matter... Similarly, vinegar only gets to wine by going first through water.'

coupling that determines so much of all that ensues in the production of form and its practices.

Above: Piero della Francesca, *Madonna del Parto*, detail, Santa Maria a Nomentana, Monterchi, after 1467.
Right: Liz Nine Months Pregnant. 1991
(Credit: Katrina Lithgow)

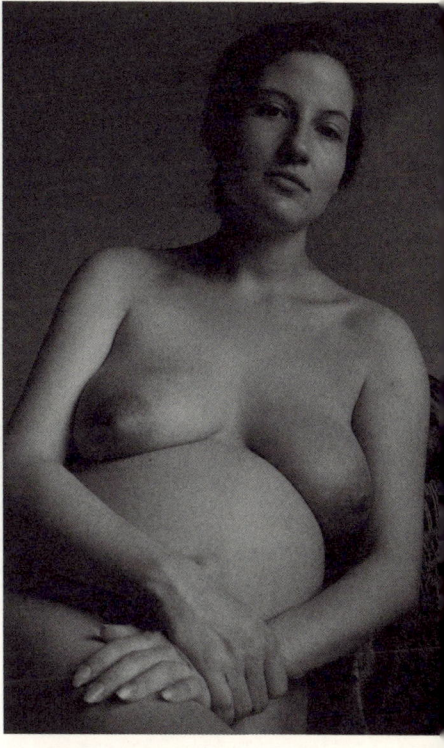

There are two images that I keep returning to in my thinking about the cleaving relations between form and matter, and the threat of interruption (read access) to this space that the gestating body poses.[21] Their connection here is both blissfully dumb (literally: they are of pregnant women) and very complex. But there is more, there is a tension between the images that though difficult to pin down, is acute.

---

21 The term cleavage presents a curious doubling of sense: the act of cleaving is simultaneously defined as both to divide *and* to connect or join. If we turn to the literal, to the face value of this term, always a material moment that, as Ingraham reminds us in 'Losing it in Architecture' (see note 8), only takes us straight back to form, we find a complex topography: cleavage is not about an *either–or* condition, but refers to the topography of the *both–and*. Literally.

Both women are in the last weeks of pregnancy – conclusive separation is imminent. Both present the viewer with a topographical cleavage of sorts. In one the cleavage is literally interrupted by the pregnancy: the rising belly splays the newly heavy breasts apart. In the other, the hand of the woman hovers over a gaping opening in her dress at the level of her swelling belly (a kind of vertical caesarean). The laces that would close the dress are generously undone, the eyelets running down either side tell us that a previous closure has been reopened: her wound is unstitched. The interruption here is chronological. Both cleavages are complex – literally interrupted by pregnancy, they no longer present the relatively simple relations of dividing and joining, or opening and closing. Instead topographical and chronological conditions are transformed. In both images, that is, pregnancy marks an interruption in the cleaving relations between form and matter.

The enduring currency of the metaphor of gestation in accounts of architectural production (the concept sketch, project development with or without abortive work, then project delivery or, bizarrely, erection) can only be explained by its crucial role dominance, to the point of disappearance, of matter by form. Within this schema, it is the role of the feminine, and within the feminine, the mother in particular, to hold or suture together incompatible orders. The feat of metaphoric structures is of course their power of substitution. Metaphors allow us to say that anything can stand in for any other thing. They operate, that is, in terms of the purely formal, existing uninterrupted by the presence of matter. But what happens when the thing that needs standing in for is itself outside of the formal, outside of representation? What happens when the thing that needs standing in for is matter itself? Nothing can stand in for matter (as standing in for is an essentially formal activity); this, if nothing else, defines matter. While error and (as a consequence) the ugly have oblique relations to matter, they cannot complete its substitution. Metaphor is all about form and matter defies formal containment, including metaphoric containment. Hence the innate and fundamental hostility that exists between matter and metaphor. Matter by definition undoes metaphor, therefore metaphors, including metaphors of matter, have to repress matter. *This* is where woman, and in particular the pregnant woman, gets caught up, as metaphoric mother and actual matrix, she is standing in for a metaphor of matter – an inherently paradoxical position.

Returning to the two images, what does this spatial and temporal interruption of the relations between form and matter mean? What is this newly plural cleavage revealing that has up to now been concealed, and that will conceal itself again? The interruption of the cleavage that describes form/matter relations constitutes, as all interruptions do, a crisis.

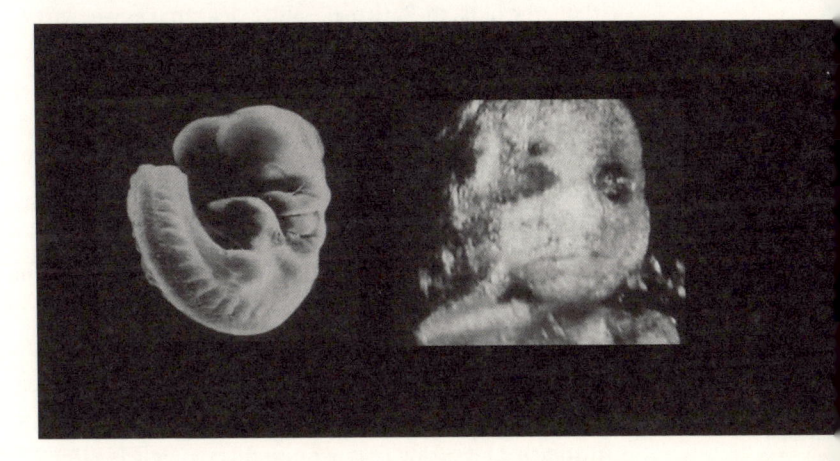

Previously reliable categories collapse. A wholly different topography emerges of proliferating cleavages that reconfigure themselves with every move, two-way, three-way, even four-way junctions. As if the body is anticipating the surrendering of form to matter that necessarily is childbirth. A surrendering that simultaneously brings about a collapse of Cartesian space, of dimension and distance, and alerts one yet again to how all these categories shore each other up. How, with the literal flood that is childbirth, this retaining construction itself must also succumb, momentarily, to the flood of matter.

The warm baby and the afterbirth are simply too large, it is inconceivable that they ever fitted in that taut belly, the dimensions do not add up. Nor does the mass lost concur with the apparent volume of fluid and flesh before me. Not only do solids and voids not add up, mass too does not. As if sublimated, matter roared in and then disappeared. I remember iodine crystals in my hot hands becoming purple wisps of vapour, curling into nothingness. I reel at my newly revealed blindness, ignorance and at the petty, crude, even pompous, techniques of my trade. I reel also at my daughter returning my gaze.

Now that I am a mother of a daughter, and not just a daughter of a mother, I have new respect for Russian dolls, a construction that neatly points to the infinity contained within as well as that contained without. The Russian doll suggests that the inside may be as big as, if not even bigger than, the outside. Both make quite clear that as we scurry around on the surface of vessels with our tape measurers we are mostly wasting our time. The immeasurable is not necessarily immeasurable because it is excessively big (or small) but because we do not know how to measure it, or more precisely, that measuring it is simply not the point. Measuring, instead, is almost the symptom of a crisis that all that is immeasurable incites. I say 'almost' because, of course, measuring does work; we can even measure how much it works. But the point here, that is compelling in regards to form and matter relations, is that measuring also doesn't work. The moment when form is unstable and matter risks surfacing is one such instance. Measuring fails because this struggle is a thing not to be seen in the first place.[22] Like the moment of childbirth, it is outside of representation. The crisis for representation (the whole business of standing in for) that it points to, is gripping. I, at least, am gripped by it. Not least because it denotes the moment of the search (the search for what the next form is to be), when form is hesitating, wavering. *This* is the opportunity to catch form off guard, on its back foot, and glimpse matter.

Extraordinary new imaging technology is currently being applied to pre-fetal monitoring. In the embryonic state, still

---

[22] The moment when matter risks surfacing is of course monstrous and, like the monsters of evolutionary theory (the redundancy of half-arrived organs, eyes on the inside, flightless wings, etc.), Darwin's insistence on gradualism, the fact that all evolution is slow and incremental, requires the monsters of embryology to evade or gaze. Or does our gaze evade them? Indeed, the problematic configu-ration of form/ function relations (what on earth does one *do* with eyes on the inside?) in embryology and evolutionary theory, that results directly from the acrobatics employed in the subjugation of matter, is remarkably similar.

suspended in the yolk sack (the umbilicus is yet to be formed) function is deferred. *This* is luxury. There is nothing (and everything) to do. We are able to watch the passing of evolutionary memory gills, a tail, a reptilian curved spine. As the relations between form and matter are played out, memory of matter that came before, surfaces as form, both redundant and extraordinary.

But there is more, there is something important here about the construction of these images. Ultrasound is used to construct stills, that when sufficiently stacked up are able to generate three-dimensional images. As if, through stacking enough photographs, a thickness is achieved that stands in for

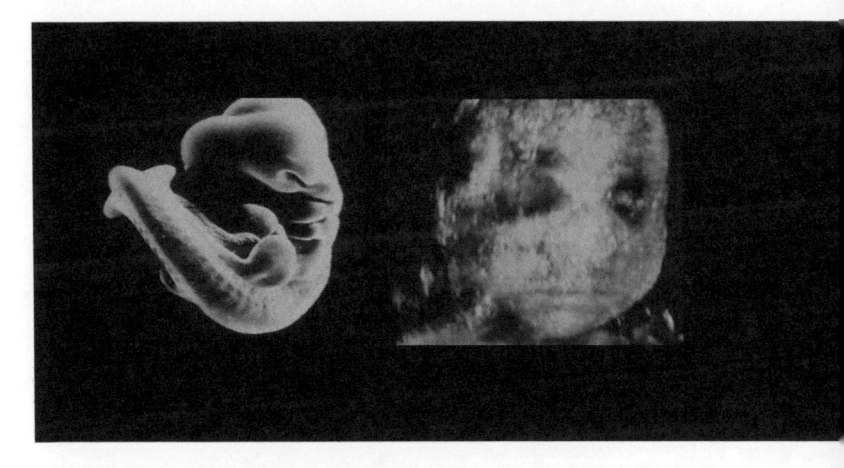

not only depth but stuff too: technology is homing in on matter. The stacking of the ultrasound stills, like the stacking of the plan drawings of the house, makes a kind of extrapolatory ham sandwich where the bread is the known, the stills, the drawings, and the ham is the gap, the error, the fudged, the lurking territory of matter. Like the belly in the belly of the Russian doll, representation, in its circling of matter, gets closer and closer and closer, but always in the knowledge that this teleology does not end with arrival.

Chapter 17

Jennifer Bloomer

Elisabeth Grosz

THE UNBEARABLE
BEING OF LIGHTNI

Chapter 17   Jennifer Bloomer
_____

Allan Janik and Stephen Toulmin

Diana Agrest

ESS

> If we cling to objects, we should trust our own clinging impulse; and once we trust that impulse we will acknowledge that such objects are precious; and once we articulate why they are precious, it will be self-evident why our desire for them must be regulated and why their benefits must be equitably distributed throughout the world. It is by crediting them that we will reach the insight that we only pretend to reach when we discredit them.
>
> Elaine Scarry, *The Body in Pain*

From that early frontier marker of modern architecture, the Crystal Palace of 1851, to any example of contemporary formalism, there can be read a strong strain, a straining edge of desire for the complete eclipse of 'matter' by 'form': form as a reification of force diagrams, form as abstraction, form – as much as is possible – without matter. Form that is, therefore, not subject to the force of gravity. Here, I am, as the architects were, thinking about a virtual lightness; after all, all that glass of light, modern and contemporary architecture, all that sand and silicon scraped from the Earth, is weighty indeed. When we speak of lightness, we refer to relations of thickness and thinness and to a building's visual relation to the earth and sky. The earth is both heavy and light, in Nietzsche's terms, subject to both its own gravity – weighed down – and that of the sun – made to fly through space in lightness. Thinness and thickness represent, or correspond to, immateriality and corporeality, which correspond to zero gravity and gravitational pull. The challenge to make the groundedness, the weightiness of architecture disappear – to make it 'light' – is an ostensibly perverse move, and an interesting one to examine.

The planet Earth is an infinitesimally minute object in an unfathomably great aggregation held together by invisible forces, the sublimity of which is so overwhelming an experience as to cast the mind into a state of pure corporeality: that moment when the mind experiences itself as a whirling, dizzy ball of matter, spinning around with all the other Big Bang detritus, nothing but a thinking ball of stuff. (But let us get back to Mother Earth.) The earth is an aggregation of particles so small that we cannot even see them, held together as if one incomprehensibly large ball by a force that we have never not known. It is a force so ubiquitously present that we are hardly aware of it, this force that holds us down and keeps us from whirling off into the light-pricked, endless black, the ethereal bepuffed blue, the fascinating no-place that has attracted and repelled humankind as long as we have existed. The force is gravity, 'being, for any two sufficiently massive bodies, directly proportional to the product of their masses and inversely proportional to the square of the distance between them'. Gravity comes to English from the Latin *gravis*, meaning heavy or serious. *Gravis* also begat the English word 'gravid', and thus, 'gravidity', which refers to the condition of pregnancy. Gravity is a force between two bodies, between aggregations of matter. If there is matter, there is gravity; if there is no matter, there is no gravity. The closer the bodies of matter are, the stronger the force; they gravitate toward each other.

Although customarily we think of gravity being present between very large masses of matter, there is gravitation between any two bodies of matter. Therefore, all material assemblages exhibit gravitation among themselves. When I stand before a building, or before a tree, or another human being, there is gravity at work between us. In a sense, all bodies of matter, great and small, are components of one great assemblage held together by an invisible and elastic structure of gravitational forces. When one of the bodies is inside the other, as in the case, say, of an organ encased in the human material assemblage, so near as to be in geographic identity, how strong is the force of gravity? Science tells us that it is so weak as to be negligible, that it is too small to be measured. The force of gravity is fabulously weak, the weakest force known in the universe. Scientific measurement works on the phenomenon of one mass inside another at the scale of large underground deposits of minerals and oil in the earth, not at the scale of the handful of pulsing matter called the human heart.

When one of the bodies has been inside the other but is no longer, as in the case of you and your mother, how strong is the urge to escape the knowledge of this negligibly weak force?

A look at Western culture will suggest an exponential, inverse proportion to the weak force of gravity. The relation of 'gravidity' and 'gravity' is serious business, a business no less weighty than its parallel relation of *mater* and 'matter'.

To see a strong and influential argument reflecting this urge to escape uneasy knowledge of the weak force (which, Sir Isaac Newton not yet having slid out of his mother's womb, had not been given its modern conception), we must look to the thinking of Aristotle. In this theoretical system, perfection, reason, natural law and form sit in opposition to the flawed or incomplete, the a-rational, chaos, and matter. Here, where a choice must be made between the unbearable weight of the body and a zero-approaching lightness, it only stands to reason (and I do not use this phrase lightly) that we would opt for lightness. This lightness represents an escape from gravity, the making of distance between oneself and the body of the other (of the earth), and even the making of distance between one's mind and one's own body, a distance-making activity delineated by Aristotle and augmented significantly for modern thought by René Descartes' early seventeenth-century discourse decisively cleaving mind and matter. From the invention of linear perspective to space exploration and rocket launches, hydroponic gardening, 'test-tube' babies, pilotis, 'light construction', to the notion of a 'City of Bits', cloning, robotics and cyborgs, we can trace the line of desired escape. The invention and popularity of virtual touring in digital space and contemporary speculation about downloading the human brain well exemplify the persistence of the urge to separate mind and body.

In an age of global capitalism, the line of escape is reinforced by post-industrial production itself. For post-industrial production necessitates an economy of material and consequently dictates a certain thinness, a certain lightness. With thinness comes the frequent need to replace decayed and obsolescent parts: the production of more manufactured bits to hold the thinness together. This is 'high technology'. Virtual reality is the ultimate capitalist dream – the ultimate economy of material, the ultimate obsolescing fact, the ultimate in a demanded, invented necessity of technological proliferation.

Furthermore, with this new mode of virtuality, an old one (television) is augmented: we have an elaborate 'switch-it-on, switch-it-off' environment, the ultimate potential in the apparent control of the object. Now you see it/feel it/embrace it; now you don't.

Software such as CATIA provides momentary answers to the problem of form, material and computers.[1] Because of the direct relation between design and production that it permits, it legitimately is touted as an extraordinary tool of material exploration and innovation in design. Any form that can be designed can be laid out in a pattern, cut and assembled, no matter how complicated the angle, no matter how irregular the curve. But there is an extreme limiting factor here, and that is the fact of material production itself. Materials – wood, metal, stone, glass, ceramic, vinyls – are imagined, and produced, to be as two-dimensional (as thin) as possible. The thinner, the less substantial materially, the more profit to the manufacturer. Thus, although the computer is imminently capable of working with wood and stone as mass, metal and glass as molten flow, this obvious possibility is blocked in the complicated nexus of production of maximum profit.[2] We are not going to escape the conundrum of capitalism in the near future. To propose otherwise is truly to create a fantasy. So... what then?

The contemporary urge to erase the distinction between the virtual and the real by rejecting the 'real' and embracing the 'virtual' is perplexing. There seem to be but two parts, or possibilities, of this erasure: (1) the virtual mimes the real; (2) the virtual displaces the real, the virtual becomes indistinguishable from the real. But what is the point after the parlour trick is discovered? This is not, in fact, a rhetorical question, because there is a point and it is a point stabbed large. And it is this: within the trick, the material world has been totally rationalised, brought under control, understood. This simulacrum with all its divergent-fast-wow possibilities

---

1   CATIA is the design software developed for the aeronautics and automobile industries and used famously in the production of the office of Frank Gehry and Associates.

2   Here I must raise the issue of global corporate capitalism as co-option of green architecture and sustainability, an investment that banks on another kind of green.

is a simple extension of the Enlightenment project as it was threaded through the modern project: ultimate lightness. The messiness and imperfections of the real, the material, i.e., the characteristics of materiality not available to us through vision, are – in theory – eliminated. But, as Elizabeth Grosz has pointed out, the real is always wrapped around the virtual, and filled with the virtual.[3] The 'real' and the 'virtual' are not the opposing concepts that the Aristotelian tradition would have them be. Grosz cites places where Nature is full of the virtual: camouflage and mimicry, for example. The 'virtual' is the space of becoming, the place where something 'new' appears, the locus of invention. And so, any virtual architectural proposal, theorised and embedded in the real, is entirely capable of becoming real, in the way that the sound coming from a radio is real, in the way that the images on television are real.

The example of bringing the material world under control by a displacement of the real with the virtual is not far removed from the abstraction of the world into a site plan in which the city is designed at 1:10,000 scale. Not a lot of material representation is possible at this scale. Nor is the detail, nor certainly any ornament. What does it mean to design at this scale? It means to think big, to 'make no small plans'.

What is at stake in the translation of a world into a diagram, into electronic impulses? What is at stake is 'Now you see it, now you don't.' (Who is it that is gone when 'Now you don't?' 'Mommy', that's who. The large-scale abstract diagram constitutes a repression of the matters of *mater* and of material, two words that are etymologically linked.) Sigmund Freud linked what he called the *fort–da* phenomenon to the young child's development of a mastery of his world. Every child at this stage, so the theory goes, loves to play a game in which he removes a desirable object from sight (*fort*), and then causes it to reappear (*da*), the latter moment accompanied by the child's great pleasure. Now you see it, now you don't. By controlling the presence and absence of the beloved object, the child expresses a real desire to control the coming and

---

3 See Elisabeth Grosz, 'Cyberspace, Virtuality, and the Real: Some Architectural Reflections', in Cynthia Davidson (ed.), *Anybody*, Cambridge, MA: MIT Press, 1997, p.108-17.

going of his mother. What giddiness comes from the ability to make mater go away and then reappear at the desired moment, in the desired form!

The repression of the matter of mater – the sensual, the material, the haptic, taste (but recall that the Latin *sapere*, the root of *sapiens* in *homo sapiens*, means 'to taste', suggesting that the tongue was apprehender of knowledge) – is expressed through our cultural identity of these things with the female. Tradition also associates the small and the detail with the female.

Architecture, 'the Mother of the Arts', is, after all (as Catherine Ingraham has pointed out), not an object art but an object-longing art.[4] The architect is rarely the builder of the objects that are so carefully projected in drawings. The building is the material object for which the architect longs when she draws. And architectural drawings, compositions of lines suggesting form, can be construed as the longing marks of architecture; or perhaps more precisely in this analogy, of the architect who is, with his conception, development, and delivery of product, a kind of mother. The mother is she who carries weight: gravidity and gravity. This relation of architect and mother is nothing new or particularly astonishing, but I raise it in order to reconsider the notion of longing – the persistent but unfulfilled desire for something – and a particular form of longing in contemporary Western architecture: *nostalgia*. The word literally means 'homesickness' (from *nostos*, return home, and *algos*, pain), and was first used in the mid-eighteenth century to describe the malady that befell sailors on ships long removed from land.

The mother of the word 'longing' is the Middle English *longen*, which means 'to seem long (to some)'. Nostalgic longing has both a temporal and a spatial dimension: it is a desiring now for a place in the past to be fulfilled in the future, accompanied by the present knowledge of the impossibility of fulfilment. Longing is slow; it seems long. Neither nostalgia nor longing has a proper place in a culture of speed and progress. In such a culture, gravity, also of interwoven

---

4   See Catherine Ingraham, 'Missing Objects', in Diana Agrest, Patricia Conway and Leslie Kanes-Weisman (eds), *The Sex of Architecture*, New York: Harry N. Abrams, Inc., p.29-40.

temporal and spatial dimensions, working at 32 feet per second squared, is a foe to be vanquished, a weak force to be overcome.

Nostalgia, gravity, matter, ornament: all of Modernism's most wanted 'criminals'. All these enemies of progress can be tracked to a single source: Mother. But Mother, the container of everyone's original home, is most especially the alarming psychoanalytic subtext of nostalgia.

When I look at the brick quoins on a neighbour's house, I am pleased by their strange embrace of its form: strong and delicate at the same time, they mark the corner as a corner is: the place where the construction is the weakest and needs reinforcement, even though this is – yes, even in the time of the house's construction, 1930 – a constructive fiction. A constructive fiction, but an architectural story – a narrative about construction that connects this house to seventeenth-century England and to sixteenth-century Italy and to many other places and times. Here the ornamental is a bedtime story handed down from mother and father to child to grandchild to great-grandchild – a story that is a history, a tradition, a structure of the human condition. Here the bricks are real; they are what they appear to be. What happens when, in 1980, the 'quoins' are translated into beefy Styrofoam extrusions sprayed with Dryvit? The detail remains trapped in abstraction; the signs of its materiality and construction are absent in this diagram. The story is still there, but it has become a story of surface; of connection to a past that itself is a reduction, an image of retro fashion-consciousness. And what does this nostalgia, this facile reference to a past that cannot be lived again, mean? Here, it is an avant-garde nostalgia: a knee-jerk attempt to move beyond the most recent past, the metal and glass box-behemoths and the concrete push-me-pull-you monsters of the prior decades, by plastering them with a thin veneer of times before.

The repression of nostalgia is at the core of the project of Modernity. But what does this mean? How is this repression manifested? (A hint: here's Mother again.) In his oft-cited essay, 'Ornament and Crime', and elsewhere, Adolf Loos identifies ornament with the barbaric and the female, and suggests that the repression of ornament is the repression of the feminine, in turn, a repression of the mother. The fact that Loos emerged from the same fin-de-siècle Viennese cultural context as Freud, the philosopher Ludwig Wittgenstein, and the journalist Karl Kraus underscores the substance of this relation.[5]

We might also cite the repression of materiality in architectural high modernism in the predominance of the white plane and the transparent plane. Further, then, the examples of Mies and Loos in the return of the repression of material in the completely sensuous materiality of the interiors of the reductive exterior forms. We might even go so far as to propose the rule of materiality as ornament in these interiors: the patterns of wood grains, the sensuous, waving and interlacing lines of polished marble. The fetishised details of Mies also constitute a site of this return-of-the-repressed.

There is also a peculiar connection of the uncanny and the female through Freud's German term for 'the uncanny': *Unheimlich*, 'the unhomely', the repressed relation between the woman who was once home and the subject's present (modern, European, urban) environment. The surrealists mined this relation in many ways. By embracing in their work the a-rational subconscious, delimited as the site of the female and the animal in the human subject, the Surrealists offered a powerful critique of the rational functionalism and machismo of National Socialism. The repression of materiality in certain strains of contemporary architecture lies on a continuum of repressions in modernism, a continuum of the embrace of masculinity and the consequent pressure for men to reject female things both within and outside themselves. The casting out of the interior from architecture proper, the division of what was once unified in architecture – interior design and architectural design – into two stereotypically gendered

---

5 The early work of Wittgenstein, and the *œuvres* of Loos and Kraus have been interpreted as reactions to the excesses of late nineteenth-century Viennese culture. Recall also that the influential *Geschlecht und Charakter (Sex and Character)*, a study that identified masculinity with goodness, productivity, and morality and femininity with evil, laziness, and amorality, was written and published in 1903 in Vienna by the virulently misogynistic and anti-Semitic boy genius, Otto Weininger, who committed suicide at 23. See Allan Janik and Stephen Toulmin, *Wittgenstein's Vienna*, New York: Simon and Schuster, 1973, and Carl Schorske, *Fin-de-Siècle Vienna*, New York: Vintage, 1961.

professions, is a symptom of this repression. Thus we can map the relations of the sublime, the uncanny, the home, the womb, the *Unheimlich*, and the homesickness that is nostalgia.

Marge Piercy's science fiction novel, *He, She, and It* (1991), features a domestic environment in which house and mother have actually merged into something close to a cyborg architecture. It is 2059, corporations occupying the remaining post-global-warming land, mass have replaced nations, the great twentieth-century cities are cavernous ruins packed with rat-like scavengers of the lower classes and ruled by violent gangs, and the purpose of architecture is simply functional: as protection from the toxic environment, as provision of circulation, and as powerful barrier ensconcing the intelligent corporate upper class. The house of Malka, the grandmother of the protagonist, is a warm, enveloping collection of rooms about a rose-bowered courtyard equipped, via digital technology, with endless knowledge and a female voice that ranges from soothing to formidably stern. The house monitors temperature, toxins, dust, who is at the door, what's for dinner, the comings and goings of children, homework and cleaning, and ministers to the emotional needs of its inhabitants. The house as a strictly formal and material environment is as familiar and comfortable as an old shoe, or, dare I say, Mom?

Piercy eschews the conventional tendency to project domestic architecture into futuristic form. The building does not take on new form; its cottage form simply becomes lifelike, an immobile, loving and caring cyborg, the perfect mother. And this womanly container liberates the fleshly women from the guilt-enlaced strictures of traditional roles for intellectual, creative and adventurous pursuits. In the novel, these pursuits take place primarily in the space of the Net, the virtual world reached simply by plugging a nipple-like apparatus of the house into a surgically implanted connection above one's ear. The real world ruin of modernity has become a *tabula rasa* on which material environments of pure exigency house bodies now obsolete as they plug into the vibrant, replacement world of electronic space, a space that has everything to satisfy any desire – for knowledge, for pleasure, for adventure, for terror, for beauty, etc. – except for the nostalgic desire for home. At the point of unplugging, the consciousness of each character in the novel returns to the mother-house, where her body has throughout been safely nestled.

In the non-fictional space-time of 1999, as well, the urge to virtual realities of any kind relies on a constant domestic space, whether proximal or distant. The space of domesticity, configured as real space, remains the spatial envelope of the cyber-venturing subject who explores the public space of the Internet or the virtual space of simulation. With his body, that hunk of pulsing meat, in his comfortable, safe, warm, uninterrupted, timeless space, he can project himself anywhere, into anything. Here, the lines of nerves and the lines of communication form a continuum. It is all transmission of information. Here, is an apparent triumph of form over matter, of the rational over the corporeal. With the ostensible obviation – secretion – of the body comes the repression of shame, sentiment, nostalgia, longing. This space of no gravity replicates in certain ways the space of the infant, or even that of the fetus: interactive intake, no responsibility to any body. A nostalgic and sentimental, if not shameful, project in the extreme: the return to the natal home, that dirty place, the matter of mater. The relentless drive toward the New is a strangely directed attempt to escape from Materia, the old, generative soil, the origin. The New is never dirty; it is always bright, spanking clean, light, full of promise, devoid of weight.

In the bottom of a battered pasteboard box that has moved with me from house to house over thirty years is a small square of glass, blue-green, about a half inch on a side. Holding it, I revisit my first trip to New York City at age 12, and especially the walk through the dizzying canyons of Manhattan. The look straight up took my breath away, but it was the automats that captivated. Wall of tiny cubicles, each with its own little glass door behind which was an edible item, the automat was Lewis Carroll come to life. Below the doors, the foundation wall was paved in colourful mosaic. I took a piece. Tiny bit of matter, it marks an unfathomably large experience. Representing little, but signifying much.

What is the status of the souvenir or the memento, with respect to nostalgia? The materiality of the memento allows it to perform as a little marker – like a birth mark or a navel – that, looked upon, calls to mind the place from whence one has come, where one has been, or, perhaps, where one wishes to return. The house, or any domestic architecture, is an aggregation of souvenirs and mementos, a container of memory – heavy matter – that stubbornly remains constant over time despite the best attempts of architects and futurists to make it otherwise.

To be nostalgic is to be pulled, whether in space or in time, toward something one has at one time experienced but cannot ever have again, a sickness for the familiar, the home. But there is a universal home that none of us can ever really go back to. That is the natal home, what Freud called the original *Heim*, and about which he went on to construct a theory of the *unheimlich*, the unhomely or uncanny, the characteristic that has been used over and over again to describe the modern Western human condition. The modern project was, in its origins, a reaction to the *unheimlich* condition of a world at war, a poisoned and polluted world, a world of little solace. It is at once the expression of the desire to escape the unhomely home of the planet and its gravity and the expression of the desire to return to a homelier home – the pure object of nostalgia and the carefree place of complete buoyancy and lightness. The project of modernity is resoundingly a project of nostalgia.

The houses in which we live are as much accretions of detail, collections of objects – of pleasures, mementos, self-projection, desire, a big mass of material matters – as they are objects that can be drawn, projected by pencil or mouse. The city, too, can be imagined as the house of culture, even and perhaps especially a multiplicitous culture, rather than a space of compositional projection and object formation: a great accretion of pleasures, mementos, desire, projections of multiplicitous and complex selves all tied to bits and pieces of matter.

All tours and all detours away from origins exist in reference to the idea of home. No matter how far one ventures into the geographical or chronological distance, there is at every point or moment the possibility of a loop in the itinerary that returns to the starting point. This home base, this safe domestic space, is an implicit, but necessary, condition of the tour that parallels that of the cyber-venturer, who can always loop back to 'shut down'. The notion of home, in all its possible iterations – from gravid vessel to the modern and contemporary space of the uncanny, to the place where your body goes to sleep at night, to the planet whose gravity we desire to escape – is implicated in this now complex heuristic apparatus that describes a territory, a space. It is an intricate system of many interconnected and interwoven parts ever changing in relation to one another and held together by a weak force. It is a strangely familiar, even homely place: a universe, a world, a nation, a city, a garden, a building, the very body within which the sentient being reading these words is inscribed.

The idea of opposing the organic to the technological is logically obsolete at the close of the twentieth century. And so, when seeking a way to rescue the significance of the material, of the pastness of the past, of aggregations of the small, we must look to models that lie outside of or over the bipolar structures set in place by Aristotle: models of materiality that are lodged deeply in the technological, but in the realm of the technological where it falls into identity (as do all opposites) with the organic; models that encompass the aggregation of the small, the collection of details, rather than simply the big, light, overall form; models supported by constructive logics; perhaps a model of working within the virtual space of the computer as if it were (as it is) a complex system of interconnected and interworking parts, the gravitational fields of which dance with each other to form a space both familiar and *Unheimlich*.

What is this place of overlapping of the technological and organic? It is something like the sentient architecture described by Piercy. Home is where the heart is, where you put your heavy little pieces of stuff. There's no place like home.

For nearly as long as human thought has been recorded, the biological has been symbolically interpreted. The consequences of this – the division of all humankind into two discrete genders and the intricate cultural constructs arising from this simple one – have been large and widespread. An important generative moment of such interpretation resides in Aristotle's influential explanation of the relation between form and matter, in which he not only associates form with the male and masculinity and matter with the female and femininity, but also represents the relation between the two, form and matter, as a kind of coitus based on the contemporaneous understanding of sexual reproduction. In *coitus* and reproduction for Aristotle, the male is the active principle, implanting its seed of complete human potential, a tiny and fully formed hominid within the germ, into a passive female, who serves only as a gestation vessel. And so it is with form and matter. The builder/architect, he asserts, plants his fully developed conception of form into inert matter, from which coitus springs the artefact. Form is ethereal, light, defined. Matter is heavy and dull, a blob of passive stuff completely in thrall to gravity, waiting, waiting to be rescued – elevated and enlightened – by Form.

Aristotle's simple model of paired binary opposites that, in their proliferation, give a biologically symbolic structure to much of Western thought and culture, persists even at the close of the twentieth century, when science has offered a plenitude of evidence that physics, chemistry, and biology do not work this way. Religion, the law, and pedagogy still rely on this model of good and evil, guilty and not guilty, right and wrong. But perhaps most perplexing of all, the discipline and profession of architecture – the art and science of building – continue doggedly to fit their intricate structure of multiple intersections onto the simple, linear binary model. For example, think of the persistence of the models of design that privilege a direct relation between originating concept and final product – form as an embodiment of idea – or that maintain clear distinctions between structure and ornament. Even, and perhaps especially, the work of those who strive to stay on the 'cutting edge' of architecture, the contemporary architectural avant-garde, work characterised by the lightest, smoothest, slickest, most folded, electronically generated forms, remains hung up, however bizarrely, on the same old binary frame. There is deep repression here.

From Le Corbusier's elevation of buildings on slender pilotis to the floating city and space colony proposals of the 1960s, to the recent phenomenon of 'Light Construction', to current electronic and virtual architectures, the desire of architecture in the twentieth century can be characterised as an expression of longing for the absence of weight, for ever more lightness. The move to lightness has a particular relation to space: lightness aims to occupy, or to give the appearance of occupying, less space. But lightness also carries an important association to time. We can see this relation in images from the late 1950s and early 1960s depicting Man soaring into the future from the present on his allegorical or physical spaceship. The present is heavy baggage, the past an unfathomable burden, but the future floats ethereally above and beyond, reachable if the technology is right, and light, enough.

To go ceaselessly forward in time is to escape the heavy, animal gravity of present and past. To go back in time is undesirable. Why? What does it mean for the human memory to go back in time as far as is possible? It is to go back into the fleshy, gravid container of one's mother's body. This vessel is thick and corpulent, but also the site of complete floating, fetal lightness, the last environment in which we experienced no gravitational pull. We could argue, then, that the move toward lightness and its concomitant escape from corporeal weight, its escape from gravity, is, at the same moment that it represents a desire to escape from the Mother (Earth's) body, also an expression of a longing for the same, or similar, site. For wherever before did we know lightness so completely and so well as within the *mater*ial container?

Chapter 18

Sadie Plant

LEARNING AND BUI
IN THE FEMININE

Luce Irigaray

Chapter 18 Sadie Plant

LDING

In the past twenty years, cybernetic systems and electronic communication networks have posed fascinating theoretical and practical challenges to orthodox Western conceptions of space, structure and identity. For feminist theory in particular, they can provide an unexpectedly rich set of resources, with the potential to do more than centuries of philosophical critique to undermine some of the most cherished and substantial patriarchal ideals.

It was in Paris in the 1970s that feminist philosophers began to ask just how far patriarchal structures extended beyond the construction of social relationships and discursive representations. In 'The Mechanics of Fluids', one of the texts in *This Sex Which Is Not One*, Luce Irigaray raised the possibility that the control of women and the representation of the feminine were continuous with some extremely basic principles of logic, of the most basic presuppositions about the nature of matter itself.[1] Western scientific and philosophical discourse has always given priority to solids and solidity, paying no attention to the ways in which fluids and fluidity work and behave. There is only a 'mechanics of near-solids'. The fluid has no terms of its own, but is simple variation on the solid theme.

Irigaray is often criticised for coming too close to an essentialist equation of the feminine and fluidity. But in this essay, the position was rather more subtle and elusive. While it is historically true that 'the properties of fluids have been abandoned to the feminine', there is, she suggested, no necessary connection between the feminine, the female or women, and the persistence of fluid matter and its disordered and informal patterns of behaviour. Nevertheless, they have been thrown together for so long that the fate of women has become inextricable from what Irigaray calls the 'Western failure to recognize a specific economy of fluids'. The social and discursive exclusion of woman is intimately connected with the inability of modern science to deal with the working of fluid, formless matter which, according to the founding fathers of the Western world, needs man and his ideas to give it shape.

---

1   See Luce Irigaray, *This Sex Which Is Not One*, Ithaca, NY: Cornell University Press, 1985, translated by Catherine Porter and Carolyn Burke from the original French edition, *Ce Sexe qui n'en est pas un*, Paris: Editions de Minuit, 1977.

These observations were astute, but they also seemed to introduce a note of extreme pessimism to the debate. If patriarchal structures ran so deep, even to distinctions between 'matter' and 'form' and the traditional prioritising of the transcendent and the ideal, how could they ever be contested? But in the past twenty years, something like the *mechanics of fluids* began to emerge from a most unlikely source, the computer, first as the theory of cybernetics, and more recently with ideas around c and connectionism. For feminist analyses of space, architecture, and the built environment, their ramifications of these new perspectives are immense. The idea of cybernetic space – cyberspace – has changed our ideas about space itself: the notion of virtuality has introduced new possibilities of undermining many old distinctions between the natural and the artificial, the organic and the inorganic, the authentic and the fake. And perhaps most significantly the proliferation of self-organising, flattened, open systems, has enabled us to move beyond the old assumption that the systems and structures need a central, organising, governing point, that something external and transcendent has to be in control.

These developments have already had a profound effect on the biological sciences, evolutionary theory, economics, the study of machine intelligence and artificial life, and a wide range of very different disciplines. And now, that the sciences have begun to pay attention to the fluids they neglected for so long, it seems that, as Irigaray suspected, they do have a dynamic of their own. The patterns of behaviour they display have leaked across the disciplines, demonstrating the same patterns and tendencies at work in what were once completely diverse areas.

It is difficult to say where this new interest in fluid dynamics, chaotic systems, and distributed networks began, but one of the most revealing places to observe their emergence is research into artificial intelligence (AI). The first wave of research in this area was conducted in the most orthodox and disciplined of terms. AI meant the expert system: a machine with an encyclopaedic knowledge, which could emulate the functions of the human expert. It took the human expert as its blueprint, and implicitly equated intelligence with the simple acquisition and rehearsal of already existing information. And it worked with serial computers, machines with central processing units and the simple ability to make one calculation at a time, albeit at an impressive speed.

But this kind of artificial intelligence soon began to reach its own limits. It was quickly discovered that machine intelligence was not a matter of the sheer quantities of knowledge in the computer, but rather a question of the machine's abilities to learn, and also to learn how to learn. Machines would only be intelligent if they were given the capacity to learn and then left to learn for themselves.

This also entailed a move away from the serial processors of the early AI research. A system, which could learn, was not a structure but a fluid, malleable network. It was not an empty vessel waiting to be filled with the contents of the human brain but a network which would evolve and grow as it learned and learned to learn – in its own way. Neural networks function as parallel processors in which multiple interconnected units operate simultaneously. Information is not stored in a single place but is instead distributed throughout the system, which makes it difficult, if not impossible, to define the state of such systems at any given time. The interconnectedness of such systems also means that subtle shifts in activity in one area can have great implications for others, again without reference to some central site. They are always in process, continually engineering themselves, growing, emerging and taking shape as a consequence of their own distributed activities. They begin to become self-organising.

In the 1980s, this kind of connectivity began to emerge with the Net, which can also be described as a parallel, distributed system which not only functions without centralised control, but also developed as a consequence of piecemeal, localised activities which have built the systems from the bottom up. Neither its growth nor its present functioning depend on the presence of some central processing unit: the Net effectively pulls itself up by its own bootstraps.

It is already the case that research in microbiology, machine intelligence and economic systems owes a great deal to this connectionist model which marks a shift away from the notion that evolutionary processes and intelligence require some central, organising, transcendent point. Even evolution begins to appear as a discontinuous series of accidents with no guiding hand, a runaway process which directs itself; a process which can be seen at work in both the organic and the inorganic worlds; a continual process of learning without a predetermined goal, without an end point, without reason. This raises the possibility of a self-organising materiality, and ultimately contests the notion that matter needs form: precisely the move of which Irigaray dreamt in her 'Mechanics of Fluids'. It is a move away from transcendence, a rejection of the idealist assumptions and the hierarchical paradigms that underwrite so much Western philosophy and cultural life. A new understanding of immanence. The learning process is the life and activity of all systems, regardless of their shape, scale and functions.

These shifts in the architecture of computer systems raise some important questions about education, teaching and learning, the production and distribution of information, and the disciplines, academic knowledge and its students.

Many of today's academic institutions resemble the early phase of artificial intelligence research, prioritising the paternal function of the professorial figure who hands his knowledge down through the generations and still referring to Platonic conceptions of knowledge, teaching, and the teacher–student relationship, all of which are based on models which prioritise the teaching of what is already known, rather than the learning of the new. And a new connectionist paradigm has enormous implications for all conceptions of teaching, learning, and the disciplining of knowledge. It brings both the chance and the need to release learning from teaching, so that teachers no longer control the acquisition of knowledge one step at a time, serial fashion, instead making it possible for students to learn for themselves. Education becomes a matter of providing resources rather than dictating information, and this in turn raises the possibility of an interdisciplinarity which does not simply collect snippets of knowledge but instead allows for a kind of decompartmentalised learning which is attentive to the same abstract process and patterns regardless of whether they are expressed in organic or inorganic, actual or virtual, real or simulated materials and environments: a connectionist approach for an increasingly interconnected world.

These reflections on the implications of the architecture of machine intelligence also raise such fascinating questions for architecture itself. As many of the speakers at *Alterities* made clear, architecture is still practised within overwhelmingly male-dominated institutions, and continues to be underwritten by some profoundly patriarchal philosophical and cultural assumptions which continue to permit the architect to play God as he draws up the blueprints and shapes the environment according to his plans.

But the top-down procedures of traditional architecture continually find themselves colliding with the complex, unpredictable realities of the material world. Twentieth-century urban planning provides one of the best examples of this tendency. Its idealised cities made beautiful blueprints but they also ran into the problem that cities are not objects of knowledge, things to be planned and designed in advance, but immensely intricate interplays of sources, interests, perspectives, zones and desires too complex and fluid for even those who inhabit them to understand.

The possibility that such complex, multiplicitous zones could ever be ruled has always been improbable. Even the most open-ended plans can go awry when, in defiance of blueprints, all those unpredictable and additional features which don't look great on paper start to appear. Weeds and grasses lift the paving stones: drugs, sex, and rock 'n' roll make their presence felt. This is not only because the movements and desires of its inhabitants cannot be captured by a city plan, but also because a city is simply not the kind of thing which can be planned. It is not a structure, but a culture, an open and dynamic system whose complexity bestows on it a life of its own. It is more akin to an eco-system than an object of knowledge to be programmed and designed. It is a complex assemblage, an intricate interplay of forces, interests, trends and tendencies too fluid for even those who inhabit them to get a grip on the whole thing. Rather than exercising any real control, plans and planners have merely added to the cacophony.

If cities display self-organising tendencies of their own, the architect may still find it possible to impose his will on smaller built environments and individual buildings. It is relatively easy to consider the city as an open, complex system – the sheer number and variety of elements which are concentrated in urban space – people, materials, traffic, energy, information, goods, waste products, and so on – make it very plausible to see the city as a dynamic system rather than a fixed structure that can be designed.

But things are changing fast at much smaller scales as well. Even buildings are beginning to present themselves as lively systems rather than inert structures, and again it is technological development which is bringing such new perceptions into play. There is now a great deal of discussion about intelligent buildings and smart materials. While much work in this area is purely speculative, the very possibility of working with buildings and materials which can to some extent express, regulate or organise themselves has the potential to disrupt many of the more traditional conceptions of the architect and architecture. Smart materials include thermoplastics and optical fibres, piezo-electrics, which respond to changes in electrical voltage, shape memory alloys, which remember their original shape and magneto-restrictive fluids, which can change their viscosity with magnetic fields. Their current applications are limited often to military uses or to the manufacture of novelty goods. But they have great practical potential: used in the construction of buildings in earthquake-prone zones, for example, such materials can show stresses and strain long before cracks become visible and can be far more flexible than even the most flexible of traditional materials.

This kind of molecular engineering, which invests materials with an unprecedented responsiveness and activity, is developing in parallel with the idea of intelligent buildings and environments. Rather like the city itself, the intelligent building becomes more akin to a processing plant with its own cybernetic systems and the ability to integrate energy and information flows. It can electronically respond to the people who use it, and there are even experimental buildings which respond to the moods and more subtle desires and inclinations of their users.

It is of course true that the vast majority of these developments are driven by desires for increasing control. But although they are unintended consequences such power-hungry researches, such developments do begin to give a new sense of activity to what has long been considered to be dead matter. And just as the architects of machine intelligence reached a point at which they had to accept that a thinking machine would only thrive if a certain degree of control was relinquished, so it is possible that intelligent buildings and smart materials will begin to express a dynamic of their own, a liveliness which changes or exceeds the desires of their designers. Anything small enough to do as it is told is likely to be smart enough to do other things as well. Buildings and materials may be no exception to this rule.

Even more significant is the extent to which they allow us to see the activity that was always immanent to matter itself. As self-organising tendencies, molecular processes and inorganic activities become increasingly perceptible, it begins to seem that the material world was not quite as inert as the Western world once thought.

Given that the many, varied needs, desires and activities of the people who lived or worked or played in their constructions have been so often overlooked, it is hardly surprising that sites, environments, and materials should have seemed so passive and inert. But a new awareness of a more enlivened world might make a significant impact on the nature of architecture and architects themselves, introducing a chance to interact with elements of the material world which once seemed to need the formal hand of man to knock them into shape. Like teachers of the future, architects may have to relinquish some control; to find a new sympathy and engagement with the worlds in which they work; to communicate with all the processes which feed into and sustain a particular building or environment become as important as the finished product; to develop long and ongoing relationships with what were once finite completed constructions.

Such new perspectives will present some fascinating difficulties for the traditional male architect who, even at the end of the twentieth century, still seems to be investing his work with *prima donna* dreams of domination and control. And although it will take more than smart materials and the new perspectives they provoke to challenge such institutionalised attitudes, it does seem very likely that the architecture of the future will need some very different architects and practices if it is to develop the responsive, interactive, and ongoing relationships it will need to enjoy all the lively elements of the lively world: buildings, spaces, materials and of course, people themselves.

It once seemed easy to demonstrate the inert passivity of matter and the random, formless nature of its processes – a world in which it could be argued that the transcendent hand of God, man or some other figure of authority was necessary to give it shape and form. If, as Irigaray argued, patriarchy thrived in such a world, it may be that its dreams of domination have made it the unwitting architect of its own downfall. Wherever one looks, it seems that the desire to impose idealist principles on the fluid processes at work in the material world is leading to the emergence of an increasingly smart and active materiality. The desire of control has begun to meet its own impossibility.